Monographs on soil survey

General Editors
P. H. T. BECKETT
V. C. ROBERTSON
R. WEBSTER

Soil survey
for engineering

A. B. A. BRINK, T. C. PARTRIDGE, and
A. A. B. WILLIAMS

CLARENDON PRESS · OXFORD
1982

Oxford University Press, Walton Street, Oxford OX2 6DP
London Glasgow New York Toronto
Delhi Bombay Calcutta Madras Karachi
Kuala Lumpur Singapore Hong Kong Tokyo
Nairobi Dar es Salaam Cape Town
Melbourne Auckland
and associates in
Beirut Berlin Ibadan Mexico City Nicosia

Published in the United States by Oxford University Press, New York

British Library Cataloguing in Publication Data
Brink, A. B. A.
  Soil survey for engineering. – (Monographs on soil survey)
  1. Soil-surveys
  I. Title II. Partridge, T. C.
  III. Williams, A. A. B. IV. Series
  631.4'7   S591

ISBN 0-19-854537-1

Filmset by Eta Services (Typesetters) Ltd., Beccles, Suffolk
Printed in Great Britain
at the University Press, Oxford
by Eric Buckley,
Printer to the University

Preface

This handbook is written for planners or engineers, and for geologists or other earth scientists. We have tried to introduce the engineer and the geologist to each other's fields, and particularly where they overlap in engineering soil surveys. If we have succeeded in equipping either to comprehend what the other is doing, we shall have achieved our aim. We hope that a geologist armed with this handbook may be able, if it is really necessary, to conduct an engineering soil survey on his own but we hope he will do it in collaboration with an engineer.

The book is intended to be compact; so, while it is intended to be comprehensive, it is certainly not exhaustive. We have referred readers to more definitive treatments of specific topics. Some readers may not be familiar with the normal ranges of values of the quantities commonly referred to by engineers, so we have offered a series of 'rules of thumb'. However, we must stress that these rules of thumb are only guidelines and readers should not apply them indiscriminately.

We are deeply indebted to Dr P. H. T. Beckett for his patient guidance at all stages of the preparation of this book and for his exhaustive editing of the manuscript. Dr A. A. B. Williams' contribution as co-author of this text was made with the kind permission of the Director of the National Building Research Institute of the South African Council for Scientific and Industrial Research. We also thank the following persons for providing unpublished information and source material: Mr D. C. Biggs, Mr S. J. Bullock, Professor J. B. Burland, Mr J. H. de Beer, Dr Finn Jørstad, Professor L. C. King, Professor A. Komornik, Dr R. F. Legget, Dr F. Netterberg, Dr R. E. Oberholster, Mr R. Proctor-Sims, Professor P. F. Savage, Mr M. D. Sugg, Dr Ing. Terracina Fernando, Dr D. H. van der Merwe, and Mr F. von M. Wagener. Mr W. J. Falla kindly assisted with the preparation of the index.

Official aerial photographs of the Republic of South Africa are reproduced under Government Printers Copyright Authority 6515 of 6 November 1979. Figure 9.1 and Table 9.1 are reproduced with permission of the Secretary for Transport, Republic of South

Africa. We are also grateful to many individuals and organizations for permission to reproduce published material; they are named in the text.

Sandton, South Africa   ABAB
Johannesburg, South Africa   TCP
Pretoria, South Africa   AABW

August 1980

Contents

Heaving of active soils
Collapsible fabric
Sinkholes and subsidences in areas of soluble rock
Soluble salts in soils
Dispersion
Special problems in arctic regions
Extremely variable soils

Plates

(The plates follow page 180)

PLATE 1. Land facet annotations over a key area representative of a land system.

PLATE 2. The two land systems on which facet annotations have been drawn occur on the same bedrock but show markedly different relief and soils; they are separated by a bold line.

PLATE 3. Land system boundaries may be delineated on small-scale photographs or other imagery from satellites.

PLATE 4. An area may be subdivided into simple soil units by airphoto interpretation with field checking at a few points—these units are of comparable size to the land facets of Chapters 9 and 10.

PLATE 5. Interpolation of soil boundaries within a network of sampling points by airphoto interpretation provides an alternative method of preparing a soil map for more detailed planning.

PLATE 6. Stereogram prepared from conventional colour airphotos, annotated for surface soil types. Note the extensive grass burning and the association of bush with outcrops, especially bouldery dykes.

PLATE 7. (a) Preliminary annotation delineates obvious boundaries around areas of rock outcrop and alluvial soils and marks points for field checking.

(b) More detail is added after the first field visit but some boundaries still require verification, including that around the area of dark tone later found to have been produced by a veld fire (arrowed).

(c) Addition of proposed borrow areas.

PLATE 8. Stereogram prepared from false colour airphotos, annotated for surface soil types. Note the red spectral signature of healthy trees and the clear definition of discontinuous rock outcrops.

PLATE 9. Mosaic prepared from infrared thermal imagery with surface soil types annotated. Note the clear definition of the strike of shallow suboutcrops and differentiation of well-drained granular soils from poorly drained alluvium.

PLATE 10. Portion of a multi-spectral colour-composite image prepared from digital data recorded by the Landsat II satellite. The accompanying annotations show land system boundaries. Note the very clear distinction between the clay residual from mafic rocks of the OP land system and the adjoining sandy soils characterizing the granite terrain of the HK land system to the north. The land system boundaries were mapped by interpreting panchromatic aerial photographs, supplemented by field checking—there is a remarkable coincidence between these boundaries and the major changes in spectral signature on the image.

PLATE 11. Stereogram constructed from standard panchromatic aerial photographs, annotated to show different soil profile units; the accompanying table summarizes the elements of the photographic image for each unit.

1. Introduction

We are constantly reminded that the world population will increase by 50 per cent during the next two decades and that all but one-tenth of this increase will be in the developing countries. This growth, coupled with the current backlog, will necessitate the doubling of most amenities during this period. Geotechnical information in the developing countries is sparse and will have to be enormously augmented as a prerequisite to efficient development planning. In the developed countries, on the other hand, most good building sites have already been exploited and more use will have to be made of sites with less favourable soil conditions. Many structures will have to go underground.

Collecting geotechnical data (i.e. the data on soils and rocks required by planners and engineers for design and construction) is to some extent a geological or geographical exercise, and the prediction of the likely conditions at an unvisited site is almost wholly so.

The engineer has to assess the magnitude of any soil problems at his site and to devise the solution which best marries the structure to the soil. Many problems are associated with particular kinds of soil or rock; their occurrence is best identified by a geologist who is familiar with the engineer's requirements. It is also the engineering geologist's task to classify soil materials on their engineering properties and to classify soil profiles on the vertical sequence of the soil materials, with particular sets of properties, that they contain.

Engineering design depends on the combination of general theory with local geotechnical data. The primary aim of this handbook is to present procedures for describing, classifying and mapping soils for engineers who need the geotechnical data on which the choice of a site or an alignment will be based, and to provide information timeously to assist in the design of what is to be constructed. To gather these data it is necessary:

(i) to record the vertical succession of soil types (soil profile) at a number of points;
(ii) to determine the engineering properties of each soil type from a number of samples;
(iii) to delineate the lateral extent of each soil profile type.

2 Introduction

The first task is achieved by direct observation and description of the soil profile and establishes what there is at each point; this is usually the responsibility of the geologist. The second is achieved by *in situ* and laboratory tests and establishes what the soil can be used for and how it can be built on; this is the responsibility of the engineer. The third is achieved by field observation aided by the interpretation of remote sensing imagery and by geophysical survey; it is usually the responsibility of the earth scientist. The contributions of the two groups of specialists must be integrated by interactive communication. Therefore we have also attempted to equip the engineer with a little geology and the earth scientist with a little soil mechanics and engineering, so that they can discuss and plan the soil survey or examinations most appropriate to their joint problems. Inevitably each will find the other's chapters very elementary.

Local geology is an important factor in engineering design and a knowledge of it is therefore a prerequisite for any engineering soil survey. 'The geological origin of a deposit determines both its pattern of stratification and the physical properties of its constituents' (Terzaghi 1955). Accordingly, this book places strong emphasis on the *origins* of soils. The properties of rocks play a vital role in the design of certain structures (e.g. tunnels and large concrete dams), but the necessary geotechnical investigations lie beyond the scope of this text.

Chapter 2 introduces the earth scientist to soil mechanics and what the engineer needs to know about soil. Chapter 3 reminds the engineer and planner of the outlines of geology. Chapter 4 acquaints the geologist with the chief requirements of the main kinds of engineering works, while Chapter 5 adds some particular groups of problems not already considered in Chapter 2. Chapter 6 describes the field investigations from which the engineer derives the specific local data on which to apply his knowledge of theory and his own experience. The remaining chapters, on soil mapping, the dissemination of geotechnical data by means of terrain classification, and terrain data banks, describe procedures of progressively smaller cost per unit area. The sequence of topics thus progresses from theory to practice. Chapters 7 to 12 pass from intensive investigations for projects involving high risk to extensive (and less accurate) investigations for those of lower risk.

Planning the design and construction of a large earth dam

illustrates the procedure of successive approximation used to avoid more expensive investigations than are necessary:

(i) Select the site, using airphoto interpretation to assist in choosing between possible alternatives (Chapters 10, 11, and 12);
(ii) Prospect for construction materials and plan access roads, using soil engineering maps prepared by more intensive airphoto interpretation and limited field check (Chapter 9);
(iii) Prove the sources of construction materials by means of a network of

TABLE 1.1

Costs of some soil surveys conducted during the past five years

Type of survey	Total cost of soil survey (£)	Cost as a percentage of cost of structure
Investigation for elevated water tower on deep (20 m) residual soil	7000	10.00
Investigation for high rise office and shopping complex on deep (40 m) till	666 000	5.20
Investigation for high rise office and shopping complex on highly variable thickness (1 to 12 m) of residual soil	355 000	1.60
Investigation for medium sized concrete dam with rock at average depth of about 6 m	26 000	1.20
Investigation for large earth-fill dam in area of relatively shallow (3 m) soil overlying granite–gneiss	37 000	0.33
Investigation for a township of 2 ha on soil with a collapsible fabric	600	
Stability analysis for township development of approximately 250 ha underlain by soluble rocks (including gravimetric survey and drilling of anomalies)	32 000	
Investigation for a township of about 400 ha with moderately complex soil conditions	13 300	
Preparation of a soil engineering map and execution of borrow-pit and centreline surveys for a trunk road 50 km long crossing variable terrain in a semi-arid area	16 600	0.30
Preparation of corridor map for railway 860 km long crossing highly variable terrain in arid area	12 500	0.02

sampling points and laboratory tests to produce a medium-scale soil map of the source areas (Chapter 8);

(iv) Define the foundation conditions and the water-retaining capacity of the materials underlying the dam site by means of detailed investigations supplemented by *in situ* tests and by laboratory tests on closely spaced samples (Chapter 7).

Each stage of a soil investigation must provide sufficient data for the intended purpose and it is the engineer's responsibility to ensure this; however, there is no advantage in the acquisition of superfluous data at unwarranted expense.

Table 1.1 shows just how expensive soil surveys and detailed investigations can be.

In general, the costs of a geotechnical investigation are between 1 and 2 per cent of the total cost of a large engineering project and more than 10 per cent of a relatively small project or where soil conditions are complex. It must be emphasized that, although cost–benefit considerations should always be foremost in planning an engineering soil survey, safety should never be sacrificed to thrift. Close collaboration between the geologist and the engineer is therefore essential at all stages of a project.

2. The engineering properties of soil

Soil, to most people, is the uppermost layer of the ground, in which plants grow. To the engineer, soil includes all unconsolidated material down to bedrock. It may be a hundred metres deep or more, and includes any materials which can be dug with a pick— albeit with difficulty. It may include very large boulders of hard rock.

There may be a gradual transition between residual soil and the underlying bedrock, so the boundary between a very stiff soil and a very soft rock is somewhat arbitrarily equated with an unconfined compressive strength of about 700 kPa.†

The topsoil, enriched in humus, is merely the uppermost horizon of the soil profile: the engineer removes it before building on the ground, because it is usually too weak to support the load of any building, and its organic content may interfere with cement stabilization of the soil for minor roads.

Thus, soil includes materials with a wide variety of properties; of particular interest to the engineer building a structure *on* or *in* the soil are those properties which affect:

(i) the support that the soil provides for any building;
(ii) the stability of slopes, either natural or in cuttings;
(iii) the lateral pressure that the soil exerts against any structure.

Often he builds a structure *of* soil, in which event further properties can be added to the list:

(iv) its suitability as a construction material;
(v) its compaction characteristics;
(vi) its ability to retain water.

The significance of these properties will be discussed presently; methods for measuring them are discussed in Appendixes A and B.

In any major construction project the engineer must select the form of foundation design which is most appropriate to both the structure and the soil on which it is to be built. To define his

---

† Unconfined compressive strength is the maximum stress which a laterally unsupported specimen of the material can sustain before breaking up.

options, he requires information on local soil conditions. For example, a large thermal power station comprises several different types of structure: a boiler house, a turbine and generator house, a coal handling plant, cooling towers, and a circulating system for ash disposal. Each of these will impart different loadings to the soil and may also affect the soil in other ways: the boiler house may desiccate the soil and lead to concomitant volumetric changes in clays, while the cooling towers and ash disposal plant may cause wetting of the soil with consequent reduction in load carrying capacity. Turbines and generators impose a vibrating load on the soil and any resulting movements could affect the precise alignment required for their satisfactory functioning.

Similar considerations apply in the case of roads, dams, and other embankment structures where the engineer is concerned with the subgrade soils, that is the materials below the foundations, and with the availability of various natural materials for their construction. In deep excavations and cuttings, on the other hand, the emphasis is on those strength properties of the soil which determine the safe angle of slope for unsupported cuts, or, alternatively, criteria for the design of soil-retaining structures.

The engineering properties of soil are controlled by the size and mineralogical composition of its individual particles, their state of packing, the manner in which they adhere to one another, and the extent to which the *voids* or spaces between particles are filled with water. Two soils of identical grading and identical mineralogical composition may behave quite differently under load because of differences in:

(i) the stability of the aggregates (or domains) formed by the platy clay particles;
(ii) the state of packing of individual sand and silt particles and clay domains;
(iii) the manner in which the particles, or aggregates of particles, are held together by cementing agents such as silica, carbonates, or iron oxides;
(iv) their degree of saturation with water.

These differences in the *soil fabric*, i.e. the spatial arrangement and orientation of the soil particles and the nature of the voids, together with differences in the mineralogical composition and size distribution of the individual particles, influence the *stress–strain behaviour* of the soil, that is the manner and extent to which it responds to stress.

Thus the engineering properties of soils may be considered under three main headings:

   (i) the properties of the individual soil particles;
  (ii) the properties of the undisturbed fabric;
 (iii) the response of the soil fabric to changing stress.

PROPERTIES OF SOIL PARTICLES

The chief properties of soil particles which are of significance in engineering are:

   (i) particle size distribution (grading);
  (ii) mineralogical composition;
 (iii) specific gravity.

All of these may be determined in the laboratory on *disturbed* soil samples, i.e. samples in which the undisturbed fabric has not been retained: Appendix A describes appropriate laboratory procedures. While specific gravity is required for calculating some properties of the soil fabric, grading and composition usually provide sufficient basis for making an engineering classification of the soil. The *indicator tests* used for such classifications comprise grading analyses (Fig. 2.1), tests to determine the Atterberg limits (i.e. the liquid limit and plastic limit, from which the plasticity index may be determined), and the linear shrinkage test (p. 262).

PROPERTIES OF THE SOIL FABRIC

The chief properties of the soil fabric which are of significance in engineering are:

   (i) moisture content;
  (ii) density;
 (iii) void ratio;
 (iv) permeability.

These are determined in the laboratory on *undisturbed* samples which preserve the fabric of the soil, though moisture content may also be determined on disturbed samples.

The moisture content of a soil is partly a function of the state of packing of the individual particles and clay aggregates, and partly of the availability of water, i.e. its situation relative to water percolating from the surface or to the water table. If the soil is not

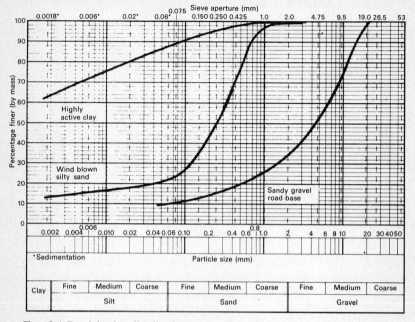

FIG. 2.1 Particle size distribution curves demonstrate the physical characteristics of soils of different origin. The highly active clay may have a PI of 50. The wind blown silty sand is non-plastic and the relatively high percentage of silt and clay shows that it is not a 'clean sand'—it is, in fact, typical of a sandy soil with a collapsible fabric. The sandy gravel, which would be ideal for a road base, is probably a 'crusher-run' material, or the well-graded product of a crushed stone plant.

saturated some of the voids will be filled with air. The percentage moisture content and the degree of saturation are both of significance in describing the soil fabric, and the laboratory procedures for determining these values are given in Appendix A. Moisture content affects almost every mechanical property of the soil with which the engineer is concerned.

The density of soil is expressed either as the bulk density γ (the weight of a unit volume of soil at natural moisture content), or as dry density γ_d (the weight of a unit volume of dry soil).† If the

---

† The units of density are in terms of weight rather than mass, since weight implies force and can thus be readily used in the calculation of pressures for engineering purposes.

moisture content is w per cent, then

$$\gamma_d = \frac{\gamma}{(1 + w/100)}.$$

Values of dry density vary from about $9\,\text{kN}\,\text{m}^{-3}$ for very highly leached residual soils to about $21\,\text{kN}\,\text{m}^{-3}$ for very highly compacted gravel. The density of undisturbed soil provides an indication of the openness of its fabric (the looseness of the packing of its particles), and hence acts as a guide to its potential compressibility. The maximum density which can be achieved by compacting a soil is a guide to its suitability as a construction material in fills and embankments.

The *void ratio e* is the ratio of the volume of voids (whether occupied by air or water) to the volume of solid particles. If V is the total volume of soil and V_s is the volume of solid particles, then

$$e = \frac{V - V_s}{V_s}.$$

The *specific volume v* is the total volume of soil containing a unit volume of soil solids:

$$v = 1 + e = V/V_s.$$

The inverse of the specific volume is referred to in practical engineering as the *per cent solid density* (e.g. when specifying that compaction for a road base of crushed stone should not be less than 86 per cent solid density).

Like other properties of the soil fabric, the void ratio is governed largely by the origin of the soil, e.g. a highly leached soil residual from granite will have a higher void ratio than an alluvial soil of similar grading. Voids smaller than $0.5\,\mu\text{m}$ are usually water-filled except in very dry soils; voids from $0.5\,\mu\text{m}$ to $30\,\mu\text{m}$ fill or empty with transpiration; voids larger than $30\,\mu\text{m}$ remain filled with water only in saturated soils, i.e. below the water table.

Permeability is the ability of the soil to transmit water through voids. The size and inter-connection of the voids, rather than the void ratio, govern the rate of seepage and hence the rate of inflow into an excavation or the leakage from a dam. A sand and a clay may have the same void ratio, but the sand has much larger inter-connecting voids and is thus highly permeable while the clay is almost entirely impermeable unless it is highly fissured. If it contains interbedded coarse and fine horizons the horizontal and vertical permeabilities of a soil may differ greatly. For this reason it is often difficult to obtain an appropriate value of soil permeability from small-scale laboratory tests (Appendix A); for design large-scale field pumping or pressure tests in boreholes are necessary (Appendix B).

Permeability affects the rate at which a soil will consolidate under load and is also an important consideration in the treatment of sub-foundation soils in dams and the prevention of seepage into excavations.

Test data which define the properties of the soil particles and the soil fabric often provide an indication of the likely response of the fabric to stress; e.g. an open fabric and a high void ratio usually indicate a highly compressible soil, while a high proportion of clay and a high moisture content are almost invariably associated with low shear strength.

STRESS–STRAIN BEHAVIOUR OF SOIL

The complexities of soil behaviour have led to the development of a number of mathematical models to explain the different patterns of response to different combinations of stress. The following very brief descriptive review attempts to provide some background to the understanding of contemporary soil mechanics theory. The engineer defines *stress* as the intensity of force acting on a surface, and the units of stress are thus force per unit area (e.g. kN m^{-2} or kPa).† He distinguishes between a *normal stress* acting at right angles to any face and a *shear stress* acting parallel to it. A normal stress forces particles closer together, or changes their state of packing or spatial distribution, i.e. it compresses the fabric. Fig. 2.2(a) illustrates the decrease in volume produced by a normal stress. The normal *linear strain* ε_z is then given by the ratio of the change in length to the original length:

$$\varepsilon_z = \frac{\delta z}{\Delta Z}.$$

† A kilopascal, or kPa, is synonymous with a kilonewton per square metre or kN m^{-2}.

Similarly the strain ε_x in the X-direction would be

$$\varepsilon_x = \frac{\delta x}{\Delta X}.$$

The proportionality factor between the strains at right angles to one another is called *Poisson's ratio* ν:

$$\nu = \frac{\varepsilon_x}{\varepsilon_z}.$$

For an incompressible material like rubber or a saturated clay which cannot drain immediately under initial application of a load,

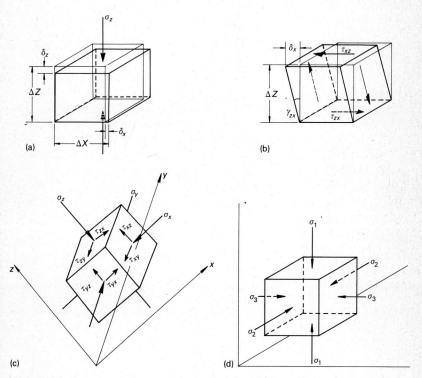

FIG. 2.2. Aspects of stress and strain in soils. (a) Normal stress and strain. (b) Shear stress and shear strain. (c) Three-dimensional stress state. (d) Principal stresses. The x, y, z axes can be transformed into any other orthogonal system, but the particular set shown in (d) refers to the principal stresses.

the theoretical value is $v = 0.5$. In non-saturated, compressible soils it may be nearer 0.3. This ratio is often assumed to remain constant and independent of the magnitude of the stress when the soil is in the elastic range.

The modulus of elasticity (or Young's modulus) E links normal stress in one direction (σ_z) with strain in the same direction (ε_z) according to the formula:

$$E = \frac{\sigma_z}{\varepsilon_z}.$$

It is often assumed that E remains independent of the magnitude of the stress in the elastic range, i.e. deformation of the soil fabric in the direction of the normal stress is linearly related to the magnitude of the stress. In short, it is assumed that the stress–strain behaviour of a soil in the elastic range is defined by Young's modulus and Poisson's ratio.

The strain produced by a shear stress is manifested in the movement of particles over each other, which changes the orientation of domains and distorts the soil fabric; it also affects the volume. The *engineers' shear strain* is the angle of distortion γ_{zx} (in radians) in Fig. 2.2(b) and is estimated by the formula

$$\gamma_{zx} = \frac{\delta x}{\Delta Z}.$$

The *pure shear strain* ε_{zx} is more convenient for some theoretical purposes:

$$\varepsilon_{zx} = \tfrac{1}{2}\gamma_{zx}.$$

The elastic *shear modulus* G is also called the modulus of rigidity and, just as E is a measure of the stiffness of the material to deformation under normal stress, so G is a measure of its stiffness to distortion under a shear stress; in Fig. 2.2(b) G is given by

$$G = \frac{\tau_{zx}}{\varepsilon_{zx}}.$$

For some purposes the elastic *bulk modulus* K is a more useful ratio: both G and K are derived from Young's modulus and Poisson's ratio as follows:

$$G = \frac{E}{2(1 + v)},$$

$$K = \frac{E}{3(1 - 2v)}.$$

(Note that if $v = 0.5$ then $G = 1/3\ E$.)

From the above discussion it will be clear that the effects of normal and shear stress are quite different and produce different kinds of strain in the fabric with corresponding changes in soil volume.

Under field conditions the soil is subjected to stresses in three dimensions. The engineer therefore considers the stresses that surround a small cube within the soil mass. He resolves the force acting on each face of this imaginary cube into one normal and two shear stresses, as shown in Fig. 2.2(c). However, it can be shown mathematically that, whatever the state of stress in the soil, it is possible to define a particular cube with its axes oriented such that no shear stresses act on any of its faces: the planes of these faces are known as the *principal planes* in the soil mass and the normal stresses acting on them are called the *principal stresses* and are designated σ_1, σ_2, and σ_3 (Fig. 2.2(d)).

The *major* and *minor* principal stresses are the largest and smallest of the three normal stresses and the difference between them is called the *deviator* stress q:

$$q = \sigma_1 - \sigma_3,$$

and is proportional to the maximum shear stress in the mass.

Finally, the *mean* stress p is the average of the three principal stresses, or

$$p = \tfrac{1}{3}(\sigma_1 + \sigma_2 + \sigma_3).$$

When the symbols refer to the interparticle stresses, rather than to total stresses, it is a widespread geotechnical convention to use the notation p', q', or σ' for what are then called 'effective stresses'. Further comments on effective stress follow later, but, like the above

$$p' = \tfrac{1}{3}(\sigma_1' + \sigma_2' + \sigma_3') \quad \text{and} \quad q' = (\sigma_1' - \sigma_3')$$

Roscoe, Schofield, and Wroth (1958) have presented an integrated theory of the stress–strain behaviour of soil. Their concept incorporates the separate idealized models of the behaviour of materials that fit the particular mechanism of soil response under different combinations of stress and void ratio. Every combination of stress and void ratio defines a particular soil state.

The integrated theory is summarized by Fig. 2.3(a) which presents the functions that link the models in terms of the mean effective stress, the deviator stress and the specific volume. All the combinations of these parameters which can occur in practice lie on or below the curved threshold surfaces depicted in Fig. 2.3(a); the intersecting threshold surfaces constitute the *state boundary surface*. As shown previously, the variables plotted in this diagram (i.e. mean effective stress p', deviator stress q', and specific volume v) are closely related to the normal stress, the shear stress and the soil fabric, respectively. The stress–strain behaviour of a soil may be considered in terms of its

(i) compressibility;
(ii) elastic deformation;
(iii) plastic deformation;
(iv) rupture.

Each of these characterizes the response of the soil fabric in a particular stress state.

For example, on the slow application of stress to a very soft soil in state A (Fig. 2.3(b)), water will be squeezed out of the voids and the fabric will be *compressed* along the line AB, i.e. the volume decreases. If the rate of loading then becomes so rapid that immediate drainage cannot occur, the soil will yield *plastically* along BC without change in volume. On reduction of the stress the fabric will not revert to its previous condition A but goes to a state represented by point D. On the other hand, application of a load to a stiff soil in state E will produce *elastic* deformation along EF. After removal of the load, the fabric produced by these high previous stresses will more or less return to its original form with small recoverable changes in volume. When a high enough load is applied to any soil its fabric will be locally destroyed by shear failure, in which different parts of the soil mass are rapidly and permanently displaced relative to one another along a shear surface. *Rupture* has occurred in the critical state at point G.

The four models of deformational behaviour for different stress

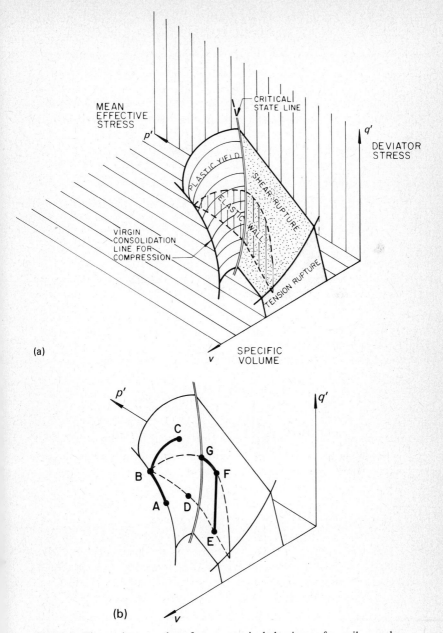

FIG. 2.3. The various modes of stress–strain behaviour of a soil may be related to various combinations of stress and void ratio in a single model, and bounded by the plastic yield surface, shear rupture surface, and tension rupture surface, which together comprise the 'state boundary surface' (a) (after Atkinson and Bransby 1978). Various stress paths within this surface are shown in (b).

states in the soil directly affect the two parameters which are commonly of most importance to the engineer, viz. compressibility and strength. Soil compressibility is a measure of the degree and rate of deformation which will occur on consolidation under a given combination of stresses. Strength is a measure of the ability of soil under severe stress to withstand rupture owing to shear failure beyond the ranges of elastic or plastic behaviour.

Compressibility

Soil is a three-phase material comprising

(i) a compressible porous skeleton formed by the fabric of the solid particles;

(ii) a virtually incompressible liquid (water) which fills some or all of the voids;

(iii) a compressible gas (air) which fills the remaining voids in any non-saturated soil above the water table.

When a load is applied to a soil mass a different stress will result in each of the three phases. The three stresses are known as:

(i) the inter-particle stress or *effective stress* σ';

(ii) the *pore-water pressure* u_w;

(iii) the *pore air pressure* u_a.

The general term pore pressure u is often used when the soil is saturated and thus contains no air. It is principally the pore pressure which governs the behaviour of soil under load. The *total stress* σ exerted on any part of the soil is equal to the sum of the effective stress σ' between soil particles and the pore pressure u:

$$\sigma = \sigma' + u.$$

The effective stress equation is written

$$\sigma' = \sigma - u.$$

This equation is fundamental to the theory of soil mechanics (Terzaghi 1936), since all measurable changes in the soil fabric are due exclusively to changes in effective stress. In the short term the principal effect of an additional load on the soil, which increases the total stress, is an increase in the pore-water pressure. This effect changes with time as water and air are expelled from the voids at a rate which is a function of their permeability. In the long term, the applied stress is transferred entirely to the solid skeleton and thereby the fabric becomes deformed. Normally the design engineer considers saturation of the soil as the worst condition that is likely to be encountered in practice. In some environments, drier conditions may be maintained and, for non-saturated soils, the effective stress equation is written:

$$\sigma' = \sigma - u_a + \chi(u_a - u_w),$$

where χ is an empirical factor that describes the proportion of the effective stress that is due to *soil suction*† $(u_a - u_w)$ (Bishop, Alpan, Blight, and Donald 1960).

The value of χ can vary between 0 and 1.0 and is akin to the degree of saturation. For a fully saturated soil $\chi = 1.0$ and the equation reverts to the simple form first given.

Compressibility may be anisotropic and vary significantly according to the direction along which stress is applied to the soil mass; e.g. the values of compressibility in directions parallel to and at right angles to a layered fabric will be different. It is therefore important that if laboratory specimens are trimmed from a large block they should be loaded in the expected direction of the applied load. Compressibility is usually measured in a consolidometer (Appendix A) and Fig. 2.4 presents the results of a single load increment causing settlement over a period. Fig. 2.5 gives the final

† Soil suction is sometimes called 'negative pore-water pressure' and expressed in engineering units, e.g. 250 kPa; it may also be expressed in terms of the pF scale, or the logarithm of the equivalent suction head measured in centimetres of water, e.g. 3.41 is equivalent to 250 kPa (Schofield 1935).

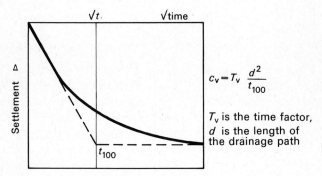

FIG. 2.4. A typical curve relating settlement and time when a fixed load is applied to a soil.

results of a consolidation test after a number of increments or decrements in load, showing the void ratio finally achieved after complete consolidation has occurred under any given effective pressure. As the effective stress increases with the application of a load, e.g. by a building or an embankment, a *normally consolidated* clay will compress along the 'virgin compression line' (Fig. 2.5(a)).

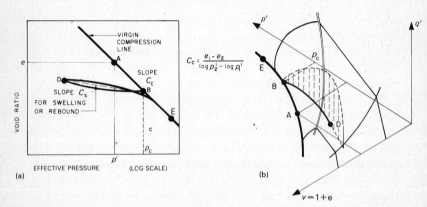

FIG. 2.5. The analysis of consolidation and rebound of a soil, including a cycle of unloading and reloading. (Points A, B, D and E refer to the same soil states in each diagram.) (a) Consolidation and rebound curves on a semi-logarithmic plot. (b) Consolidation and rebound as conceptualized in the model of Fig. 2.3.

If the load is removed from the clay a certain amount of rebound or swelling will occur. On reloading the soil again consolidation will very nearly follow the rebound path and rejoin the virgin compression line at the *pre-consolidation pressure* p_c. Any clay which behaves according to the pattern of this second cycle is described as *over-consolidated*: in this state the existing effective pressure is less than some previous maximum pressure to which the material had been subjected. In other words, the clay would be in a normally consolidated state along the *irreversible* loading paths from A to B (Fig. 2.5) and from B to E. The unloading/re-loading path from B to D may be considered *reversible*, and the clay is in an over-consolidated state along this path. The implications of these terms are important: at low stress normally consolidated soils behave plastically whereas over-consolidated soils deform elastically in the same stress range. If the stress applied to an over-consolidated clay exceeds the pre-consolidation pressure, then the consolidation will be much larger than if the stress had remained below this value.†

Appendix A describes the procedure for determining the *coefficient of consolidation* c_v (p. 271), which is a measure of the rate at which consolidation takes place, and the *compression index* C_c or *swelling index* C_s, which are measures of the amount of consolidation. Ideally the values of c_v and C_c would be calculated with the equations in Figs. 2.4 and 2.5 from the results of a consolidometer test carried out on a small intact uniform specimen in which the drainage path, i.e. the distance over which pore-water pressures will dissipate, could be assumed to be related to the thickness of the specimen. In the field, however, interconnected discontinuities in the soil fabric such as root holes, termite channels, fissures, or intercalated lenses of different texture shorten the drainage paths. The time required for the consolidation of a large soil mass may thus be a hundred or a thousand times less than that expected for a homogeneous layer of the same thickness, particularly as this time is proportional to the square of the length of the drainage path. Accelerated settlement can be induced by exploiting this relationship, for example by installing vertical sand drains in the soil beneath an embankment; if such drains are spaced 2 m apart within a compressible soil horizon 10 m thick, the drainage paths are reduced from 5 m to 1 m with a corresponding reduction in the time required for settlement from, say, 6 years to 3 months.

The settlement to be expected beneath a foundation can be estimated from the consolidation curves. The change in void ratio induced in each soil layer affected by the applied pressure can be calculated from C_c. Then the consolidation in each layer can be calculated from its thickness and the change in its void ratio or specific volume. Finally, the total consolidation that is likely to occur throughout the soil profile influenced by the foundations can be obtained. This is a measure of the total settlement of the structure. To calculate the rate of settlement from c_v it is also necessary to know the appropriate *time factor* T_v which depends on various boundary conditions that affect the development of pore

---

† The pre-consolidation pressure is of considerable significance in foundation design since, if foundation pressures are maintained below it, only relatively small settlements will occur which can be accommodated by many structures.

water pressure, and particularly on the rate of drainage possible at the boundaries of the relevant soil horizons.

Example of a settlement calculation

The calculation of settlement is complicated by the large number of variables, some of which can only be assessed by engineering judgment, so that practical predictions must be checked against the expectation and intuition of the engineer. Nevertheless, the following simplified example serves to illustrate the principles involved.

The potential consolidation settlement of a storage tank on the simple soil profile of Fig. 2.6, and the time required for 90 per cent of this settlement to take place, are to be calculated. The *initial* effective pressure p'_0 at the middle of the layer of normally consolidated clay, owing to the weight of the natural overburden, can be calculated:

$$p'_0 = (2 \times 15.7) + (1 \times 9.4) + (2.5 \times 7.9) = 60.5 \,\text{kPa}.$$

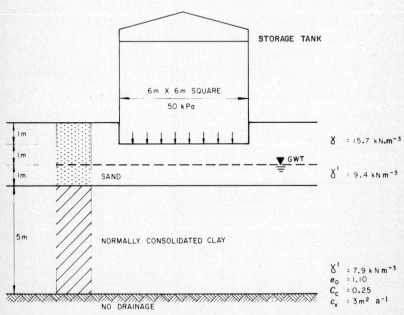

FIG. 2.6. Practical example of the calculation of consolidation settlement in clay.

The *additional* effective pressure $\Delta p'$ on the soil immediately beneath the foundation of the tank owing to the net applied loading after the removal of 1 m of sand is:

$$\Delta p' = 50 - (1 \times 15.7) = 34.3 \text{ kPa}.$$

The average increase in the vertical stress on the clay at any given depth within the zone of influence of the foundation is obtained by assuming an elastic distribution of stress with depth down to the middle of the layer in question. In this example the middle of the clay layer is 4.5 m below the foundation; the stress reduction factor appropriate to the dimensions is selected from standard tables (Winterkorn and Fang 1975) and is a function of the number, thickness and elastic moduli of the various soil horizons and the size and shape of the loaded area. In this case:

$$\Delta p' = 0.3 \times 34.3 = 10.4 \text{ kPa}.$$

By arithmetic manipulation of the equations linking the variables, the consolidation settlement ρ_c can be expressed as:

$$\rho_c = \frac{KC_c}{1 + e_0} \times H \log_{10} \left(\frac{p_0' + \Delta p'}{p_0'} \right),$$

where K is a pore pressure factor (1.0 for a normally consolidated clay), e_0 is the initial void ratio, and H is the thickness of the layer. Thus

$$\rho_c = \frac{1.0 \times 0.25}{1 + 1.10} \times 5 \log_{10} \left(\frac{60.5 + 10.4}{60.5} \right) = 0.041 \text{ m} = 41 \text{ mm}.$$

For the particular configuration of drainage conditions in this case and for a 90 per cent degree of consolidation, $T_v = 0.80$. The time required for 90 per cent of the consolidation to take place, t_{90}, is given by an equation similar to that presented in Fig. 2.4, so that

$$t_{90} = \frac{T_v d^2}{c_v},$$

where d is the length of the drainage path.

In this case it is assumed that there are no discontinuities in the clay, so that the length of the drainage path is equal to the thickness of the layer, i.e. $d = H$. Thus

$$t_{90} = \frac{0.80 \times 5^2}{3} = 6.67 \text{ years.}$$

Note that if the clay were underlain by a much more permeable layer the length of the drainage path would be half its thickness.†

Elastic behaviour

The soil behaves *elastically* when the stress state lies within the zone below the state boundary surface of Fig. 2.3(a). Such behaviour, under both normal and shear stress, is conceptualized by the insertion of an 'elastic wall' beneath the state boundary surface. This wall can be surmounted anywhere along its length at its intersection with the state boundary surface, provided that this occurs in the direction of decreasing specific volume. So, once a clay has been compressed, it can never revert to a condition in which it has a greater specific volume than is permitted by the elastic wall passing through the point which represents the maximum stress to which it has ever been subjected (i.e. the pre-consolidation pressure), unless it is dug up, loosened, or mixed with more water, i.e. unless it is remoulded.

The elastic behaviour of the soil is significant in calculations of the rebound, or upward deformation, of the floor or sides of a deep excavation due to release of stress. It is also useful in calculating the immediate settlement, due to elastic deformation, of a soil under an applied load.

Ten or twenty years ago there was little need to know the actual stress–strain parameters, because the practising engineer had no time to perform the necessary calculations with the 'manual' mathematical techniques then available. Nowadays, with computers for numerical analysis, it is possible to apply methods of elastic solution, e.g. the finite element method, and there is great need for realistic values of soil parameters and hence for an understanding of the conditions governing their variation.

---

† In an over-consolidated clay a similar procedure would be followed in calculating the likely settlement, but those portions of the settlement due to consolidation in the stress range below and above the pre-consolidation pressure should be carefully distinguished so that the appropriate compression index for each range is used (either C_s or C_c).

Plastic behaviour

Plastic deformation involving permanent distortion of the soil fabric occurs when the soil is in a stress state represented by a point on the curved yield surface in Fig. 2.3(a). On further loading, the state will traverse the curved plastic yield surface and both plastic strain (irrecoverable yield) and some elastic strain (recoverable strain) will occur; compressive strain (consolidation) may also occur but only if the soil volume can change through drainage. If the state of the soil is represented by a point below the state boundary surface it must be over-consolidated to some degree, so the path traced by any subsequent response to an increment in stress must lie on the particular elastic wall defined by its pre-consolidation pressure, until the path intersects the state boundary surface. The strains resulting from plastic behaviour under stress are calculated according to the theory of plasticity which is widely applicable to various materials, and particularly to metals in the ductile range. Plastic strain is manifested, for example, when a flexible bulkhead or sheet-pile wall, retaining soft fill, bulges when additional loads are imposed on the fill by a crane; this type of distress does not involve failure along a single shear plane.

Rupture

Rupture may occur when a soil is subjected to either shear or tensile stresses of the appropriate magnitudes (Fig. 2.3(a)). Under shear stress, rupture occurs when the maximum shearing resistance or *shear strength* of the soil is exceeded.

The value of the shear strength of a soil is used to assess its ability to support heavy loads without undergoing sudden and catastrophic deformation due to shear failure. It governs the ultimate bearing capacity† of the soil and the stability of slopes and is also used in the calculation of the earth pressure behind a retaining structure. Shear strength is the sum of the frictional resistance between the soil grains and the cohesion imparted by the finer clay

---

† *Ultimate bearing capacity* must be clearly distinguished from *safe bearing pressure*. The former is that pressure at which shear failure will occur in the soil beneath a foundation; the latter is that pressure which soil within the zone of influence of a foundation can support without causing shear failure. Further, the *allowable bearing pressure* is that pressure which the soil can support without causing shear failure or undergoing consolidation sufficient to cause settlement and distress within the structure, and this may be lower than the safe bearing pressure.

and silt fractions. This cohesion depends to a large extent on the moisture content of the soil.

The shear strength or shearing resistance may be expressed by the equation illustrated in Fig. 2.7:

$$\tau = c' + \sigma' \tan \phi'$$

where c' is the cohesion in terms of effective stress, σ' is the effective stress, and ϕ' is the angle of shearing resistance in terms of effective

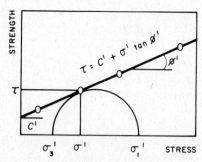

FIG. 2.7. The shearing resistance of a soil increases with the normal stress and can be represented by the two components of cohesion and friction.

stress. It should be noted that the angle of shearing resistance is sometimes referred to as the 'angle of internal friction', but this term is not advocated as it implies that there is only one value of this parameter. In clays the angle of shearing resistance depends on the type of test conducted, although in sands there is no such discrepancy. 'Angle of repose' is another term used with reference to sand and gravel, for which it is synonymous with the angle of internal friction. In cohesive materials the inclination of a stable slope also depends on the height of the slope.

Laboratory procedures for determining shear strength and unconfined compressive strength are given in Appendix A (pp. 271–4). The unconfined compressive strength is relatively simply determined and gives a rough approximation of the shear strength under confined conditions; the shear strength of an intact saturated clay under undrained conditions is half its unconfined compressive strength.

When using the equation for shearing resistance it must be remembered that the properties of the entire soil mass are controlled as much if not more by structural features such as fissures, joints, and slickensides, which affect the location of any potential plane of shear failure, as by the interactions between individual soil particles. The shear strength along a fissure in the soil mass is considerably lower than that through intact material and, in fissured clays, may be only 30 to 50 per cent of the intact strength.

Soil loses strength to varying degrees when it undergoes strain, or displacement along a plane of shear failure. The remaining or *residual strength* is not usually attained until considerable displacement has occurred. In laboratory tests on stiff clay the peak strength is achieved after a few millimetres of shear displacement (point A in Fig. 2.8), while a displacement of about 100 mm is required to generate the residual strength (point B). The reduction in strength is the result both of loss of cohesion and of reduction in frictional resistance (Fig. 2.9) owing to a change in the orientation

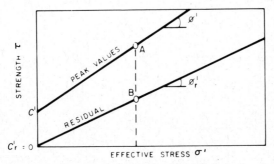

FIG. 2.8. There is a reduction in strength of a stiff clay when it undergoes strain due to a change in the orientation of clay platelets.

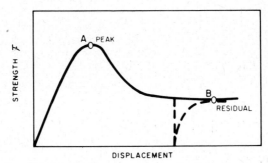

FIG. 2.9. Considerable displacement is required to produce the strain softening behaviour in a stiff clay.

of clay platelets. The points A and B in Figs. 2.8 and 2.9 are obtained from tests on clay in a shear box as described in Appendix A.†

The residual shear strength is of practical significance when assessing the stability of slopes in stiff clays, in which intact lumps may be relatively strong but pre-existing discontinuities reduce the strength of the soil mass. The discontinuities may be joints, fissures, or slickensides or other pre-existing shear planes. Joints or fissures usually have near vertical or horizontal inclinations, and their rough surfaces usually maintain a shear strength greater than the residual value. Slickensides usually form a discontinuous network of slick surfaces which are inclined closer to the horizontal than to the vertical. The shear strength of a highly slickensided clay is usually close to the residual value, so large embankments have collapsed due to 'base failure' in a thin underlying horizon of stiff slickensided clay. Cut slopes for lined canals have sometimes failed after a time owing to the release of existing stresses in the soil mass as a result of excavation; this stress release allows water to enter fissures and causes further reduction in residual shear strength. Pre-existing shear planes are frequently more continuous and exhibit striations on the shear surface. Along such a shear plane the strength of the soil mass is reduced to the residual value and, if shear stresses are increased by the imposition of an external load, or by the cutting of a steeper slope, failure will recur along the plane of weakness. This presents a hazard in old landslide zones. Table 2.1 suggests which strength parameters should be used for practical design in clays of different types and with different structural features.

Example of stability analysis for a cut in soft intact clay

The stability of a slope cut in clay, as in Fig. 2.10(a), may be assessed by considering the simple mechanism of slip along a

† Most over-consolidated or stiff clays exhibit 'strain-softening', resulting largely from particle re-orientation, or a change in fabric, within the zone of shearing. The consequence of this behaviour of stiff clays in practice is the phenomenon of 'progressive failure' whereby the propagation of failure within a mass is caused by the progressive softening of one over-stressed portion after another. The factor of safety of many excavations in stiff clay is thus likely to decrease with time. On the other hand the behaviour of soft clays and most granular materials is akin to 'strain-hardening', whereby the fabric is altered into a denser state of packing through application of a stress, and the factor of safety of an embankment or footing on these materials will increase with time.

TABLE 2.1

Strength parameters to be used in the analysis of the stability problems of slopes in clay. (Modified from Skempton and Hutchinson 1969)

Condition	Typical consistency	Structure	Type of clay	Shear strength parameters	
				Short term	Long term
First time slides	Soft	Intact	Normally consolidated	xc_u	$c'=0, \phi'$
	Firm	Intact	Lightly overconsolidated	xc_v	c', ϕ'
	Stiff	Intact	Overconsolidated	xc_u	c', ϕ'
	Stiff	Jointed	Overconsolidated	fxc_u or $c'=0, \phi'$	$c'=0, \phi'$
Repeated slides	Stiff	Slickensided	Overconsolidated	c_r', ϕ_r'	c_r', ϕ_r'
	Any category	Pre-existing shear surface in mass		c_r', ϕ_r'	c_r', ϕ_r'

c_u: peak strength parameter, undrained;
c', ϕ': peak strength parameters, drained;
c_r', ϕ_r': residual strength parameters ($c_r' \to 0$);
x: reduction factor for rate of testing, anisotropy, etc. (judged from engineering experience);
f: reduction factor for fissures (judged from engineering experience).

circular arc such as is often observed on natural slopes and in man-made cuts. The disturbing force is the weight of the mass W and this is resisted by the shear strength of the clay acting along the potential slip surface. The solution to the problem of mechanical equilibrium requires that moments are taken about the centre of the slip circle O (a 'moment' is force times distance). When the slide is imminent the maximum shear strength of the clay c_u is mobilized along the length of the potential slip surface l which has a radius R. The factor of safety F is given† by the ratio of the moment of the resisting forces M_R to the moment of the disturbing forces M_D:

$$F = \frac{M_R}{M_D} = \frac{(c_u l)R}{Wd}$$

$$= \frac{(18 \times 12) \times 7.5}{632 \times 2.5} = \frac{1620 \text{ kN m}}{1581 \text{ kN m}} = 1.02.$$

In other words the slope is stable, but only marginally so.

If the tension crack at the top of the slope now becomes filled with water, an additional disturbing moment, Ph, of 51 kN m is produced by the water pressure acting horizontally on the soil mass (Fig. 2.10(b)). Thus

$$F = \frac{1620}{1632} = 0.99,$$

i.e. the slope is now unstable and failure will occur as in Fig. 2.10(c). This illustrates the typical features of a rotational slide with a scarp exposed at the top of the slope and a 'bulge' formed at the toe.

This is a simple example, but it illustrates the engineer's approach to the problem: he first postulates a mechanism, then assigns values to the soil parameters, and finally considers the conditions of equilibrium at the limit of stability. His judgment of which mechanism is appropriate, and the validity of the soil properties, both depend very much on accurate soil profile descrip-

---

† The factor of safety F is 1.0 when a state of equilibrium is only just maintained, i.e. this is the critical value below which failure must occur. For slope stability a factor of 1.5 is considered safe; this may be reduced to 1.3 when designing to allow for sudden drawdown of water in an earth dam. In calculations of safe bearing pressure a factor of safety of 3 is used to avoid overstressing any layer within the zone of influence of the foundations.

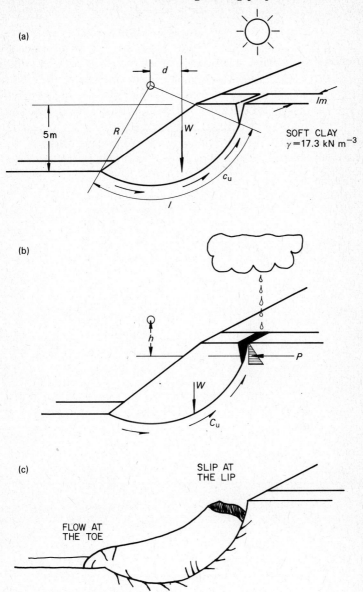

FIG. 2.10. Calculations of the stability of a slope must take account of the mechanism of failure and the forces involved. (a) Potential slip circle. (b) Effect of water pressure. (c) Landslide failure.

tion. In practice there may be many factors which complicate the picture and demand sound engineering judgement.†

A few examples are given in Fig. 2.11. The side slope in a canal might be quite stable if submerged but dewatering will remove the force U_2 while U_1 may still remain, causing a reduction in stability. A major joint in the soil or rock mass may give rise to sliding of a block on a plane of weakness rather than a circular slip. Fissured materials are prone to progressive failures and surface sloughing. A wedge failure may result from certain orientations of fissures. In a frictional material, the friction component should be included in the

† In contemporary engineering practice the methods of stability analysis are more sophisticated so as to cope with the many variables actually encountered in each different situation. In such analyses the mass is split up into a number of imaginary vertical slices to allow for variations along a potential slip surface, which need not be circular ('method of slices'), or the mass is considered to be composed of a number of blocks, or wedges, which slide on each other ('wedge analysis'). When analysing a large number of possible conditions the numerical manipulations can become very time-consuming, but many computer programs are available for judicious application by the engineer.

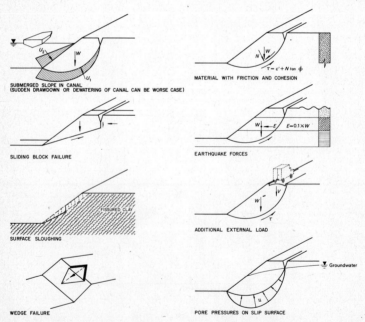

FIG. 2.11. There are many different mechanisms of failure for a slope and a variety of body forces or external loads which can produce it.

resisting forces for calculation of the factor of safety. An earthquake may introduce a sudden horizontal body force E, depending on the mass W, which may upset stability. External loads due to heavy vehicles will also increase the disturbing force. The ground water level always affects the stability of a slope and the disturbing force depends on the actual distribution of the water pressure.

COMPACTION CHARACTERISTICS

When soils are used as construction materials in any type of fill or embankment they nearly always require compaction to prevent settlement and to reduce permeability. For any type of soil there is a limit to the density which can be achieved by compaction under a given effort; this maximum density can be attained only at a particular moisture content known as the *optimum moisture content*. Below the optimum moisture content the soil fabric resists re-arrangement of the particles into a state of close-packing, whereas above it the presence of water in the voids inhibits the achievement of closer inter-granular contact. The relationship between moisture content and the density attained by compaction is shown in a *moisture/density curve* (Fig. 2.12). Compaction at the construction site is achieved by placing soil in layers with a thickness of about

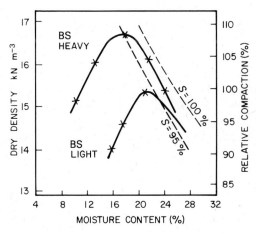

Fig. 2.12. At the optimum moisture content compaction will achieve a maximum density in a soil. S is the degree of saturation of the soil.

150 to 300 mm. The soil in each layer is brought to its optimum moisture content and then compacted by an appropriate number of passes of a roller, vibrating plate, impact compactor, or other equipment before the next layer is added.

The properties of a soil may be improved by the addition of Portland cement, slaked lime, or other *stabilizing agents*. Cement is used to increase the strength of sandy soils, while lime, which is ineffective with sands, reduces the plasticity and increases the strength of clayey soils.

The strength that a material will achieve on compaction is measured by a penetration test (Appendix A) to determine the California bearing ratio (CBR). Table 2.2 (see next section) gives typical values of CBR for various soils and their potential use as construction materials. When soils are to be stabilized with cement or lime the CBR test is sometimes supplemented by other strength tests, e.g. triaxial tests or determinations of unconfined compressive strength, which are carried out on samples of materials after they have been treated and compacted in a cylindrical mould.

CLASSIFICATION OF SOILS FOR CONSTRUCTION

When engineering works are constructed of soil the response of different soil materials to compaction and stabilization treatments is of paramount importance. Soil classification systems are often useful for predicting the compaction characteristics of a soil from previous experience. Two systems are widely applied in civil engineering. The first of these was developed for the US Corps of Engineers (Casagrande 1947) and is now known as the Unified Classification System (Table 2.2) (US Army Corps of Engineers 1953). A sound knowledge of this system will prove sufficient for classifying most soils and may assist in predicting their behaviour as construction materials. In Fig. 2.1 the clay would be a CH material, and the sand and gravel SM and GW materials, respectively. Fig. 2.13 facilitates rapid identification of the Unified class of any material.

The second system, the revised Public Roads Administration Classification or PRA classification (Allen 1945), defines soil classes specifically for road design and construction purposes (Table 2.3). In this system the parenthetic group index reflects the combined effects of the silt fraction, the liquid limit and the plasticity index on

TABLE 2.2

The Unified Classification System is a direct guide to the use of soils as construction materials

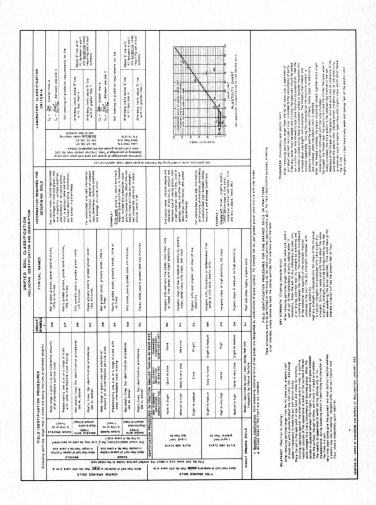

(a) Identification and description

Prepared by the United States Army Corps of Engineers in consultation with Professor A. Casagrande. Published with kind permission of the United States Department of the Interior, Water and Power Resources Service.

Table 2.2—*continued*

(b) Soil characteristics pertinent to roads and airfields

Typical name	Group symbol	Value as *subgrade* when not subject to frost action	Value as *subbase* when not subject to frost action	Value as *base* when not subject to frost action
Well graded gravels or gravel–sand mixtures, little or no fines	GW	Excellent	Excellent	Good
Poorly graded gravels or gravel–sand mixtures, little or no fines	GP	Good to excellent	Good	Fair to good
Silty gravels, gravel–sand–silt mixtures	GM	Good to excellent	Fair to good	Poor to good
Clayey gravels, gravel–sand–clay mixtures	GC	Good	Fair	Poor to not suitable
Well-graded sands or gravelly sands, little or no fines	SW	Good	Fair to good	Poor
Poorly graded sands or gravelly sands, little or no fines	SP	Fair to good	Fair	Poor to not suitable
Silty sands, sand–silt mixtures	SM	Fair to good	Poor to good	Poor to not suitable
Clayey sands, sand–clay mixtures	SC	Poor to fair	Poor	Not suitable
Inorganic silts and very fine sands, rock flour, silty or clayey fine sands or clayey silts with slight plasticity	ML	Poor to fair	Not suitable	Not suitable
Inorganic clays of low to medium plasticity, gravelly clays, sandy clays, silty clays, lean clays	CL	Poor to fair	Not suitable	Not suitable
Organic silts and organic silt-clays of low plasticity	OL	Poor	Not suitable	Not suitable
Inorganic silts, micaceous or dia-tomaceous fine sandy or silty soils, elastic silts	MH	Poor	Not suitable	Not suitable
Inorganic clays of high plasticity, fat clays	CH	Poor to fair	Not suitable	Not suitable
Organic clays of medium to high plasticity, organic silts	OH	Poor to very poor	Not suitable	Not suitable
Peat or other highly organic soils	Pt	Not suitable	Not suitable	Not suitable

Source: United States Army Corps of Engineers (1953).

Potential frost action	Compressibility and expansion	Drainage characteristics	Compaction equipment
None to very slight	Almost none	Excellent	Crawler-type tractor, rubber-tyred roller, steel-wheeled roller
None to very slight	Almost none	Excellent	Crawler-type tractor, rubber-tyred roller, steel-wheeled roller
Slight to medium	Slight to very slight	Fair to poor	Rubber-tyred roller, sheepsfoot roller
Slight to medium	Slight	Poor to practically impervious	Rubber-tyred roller, sheepsfoot roller
None to very slight	Almost none	Excellent	Crawler-type tractor, rubber-tyred roller
None to very slight	Almost none	Excellent	Crawler-type tractor, rubber-tyred roller
Slight to high	Very slight to medium	Fair to practically impervious	Rubber-tyred roller, sheepsfoot roller
Slight to high	Slight-medium	Poor to practically impervious	Rubber-tyred roller, sheepsfoot roller
Medium to very high	Slight-medium	Fair to poor	Rubber-tyred roller, sheepsfoot roller close control of moisture
Medium to high	Medium	Practically impervious	Rubber-tyred roller, sheepsfoot roller
Medium to high	Medium-high	Poor	Rubber-tyred roller, sheepsfoot roller
Medium to very high	High	Fair to poor	Sheepsfoot roller, rubber-tyred roller
Medium	High	Practically impervious	Sheepsfoot roller, rubber-tyred roller
Medium	High	Practically impervious	Sheepsfoot roller, rubber-tyred roller
Slight	Very high	Fair to poor	Compaction not practical

Table 2.2—*continued*
(c) Typical soil properties after compaction

Group symbol	Modified AASHO compaction§			Standard Proctor compaction‖			
	Max. dry density kN m^{-3} †	Optimum moisture content %	Void ratio e at optimum compaction ‡	Max. dry density kN m^{-3}	Optimum moisture content %	Void ratio e at optimum compaction	Permeability k cm s^{-1}
GW	20.0–22.5	11–8	<0.35	>19.0	<13.3	ID	$(2.7 \pm 1.3) \times 10^{-2}$
GP	17.5–22.5	14–11	<0.50	>17.6	<12.4	ID	$(6.4 \pm 3.4) \times 10^{-2}$
GM	18.5–23.0	12–8	<0.40	>18.2	<14.5	ID	$>3 \times 10^{-7}$
GC	21.0–23.0	14–9	<0.30	>18.4	<14.7	ID	$>3 \times 10^{-7}$
SW	17.5–21.0	16–9	<0.40	19.0 ± 0.8	13.3 ± 2.5	0.37 ± ID	ID
SP	16.0–21.5	21–12	<0.70	17.6 ± 0.3	12.4 ± 1.0	0.50 ± 0.03	$>1.5 \times 10^{-5}$
SM	17.0–21.5	16–11	<0.60	18.2 ± 0.15	14.5 ± 0.4	0.48 ± 0.02	$(7.5 \pm 4.8) \times 10^{-6}$
SC	16.0–21.5	10–11	<0.35	18.4 ± 0.15	14.7 ± 0.4	0.48 ± 0.01	$(3 \pm 2) \times 10^{-7}$
ML	14.5–21.0	24–12	<0.70	16.5 ± 0.15	19.2 ± 0.7	0.63 ± 0.02	$(5.9 \pm 2.3) \times 10^{-7}$
CL	14.5–21.0	24–12	<0.80	17.4 ± 0.15	17.2 ± 0.3	0.56 ± 0.01	$(8 \pm 3) \times 10^{-8}$
OL	14.5–17.0	33–21	<0.90	ID	ID	ID	ID
MH	13.0–17.0	40–20	<0.70	13.2 ± 0.65	36.3 ± 3.2	1.15 ± 0.12	$(1.6 \pm 1.0) \times 10^{-7}$
CH	13.0–17.5	36–19	<0.90	15.0 ± 0.3	25.5 ± 1.2	0.80 ± 0.04	$(5 \pm 5) \times 10^{-8}$
OH	13.0–17.5	45–21	<0.70	ID	ID	ID	ID
Pt	—	—	—	—	—	—	—

† Source: United States Army Corps of Engineers (1953).
‡ Source: Road Research Laboratory (1961).
§ Wherever we have used the term 'Modified AASHO' we imply a heavy compactive effort similar to BS 1377 of 1975 (test 13) or ASTM D 1557-78 or AASHTO test T180-74.
‖ This is closely equivalent to the light compactive effort of BS 1377 (test 12) or AASHTO T99-74.
± Indicates 90% confidence limits.
ID Indicates insufficient data.

| Standard Proctor compaction | | | | | | Typical design values | |
| Compressibility % Reduction of initial volume | | Shearing strength | | | | | |
at 137.9 kN m^{-2}	at 344.75 kN m^{-2}	c_0 kN m^{-2}	c_{sat} kN m^{-2}	Tan ϕ	ϕ (Effective stress envelope) degrees	CBR	Subgrade modulus k MN m^{-3}
< 1.4	ID	ID	ID	> 0.79	> 38	40–80	80–135
< 0.8	ID	ID	ID	> 0.74	> 37	30–60	80–135
< 1.2	< 3.0	ID	ID	> 0.67	> 34	20–60	55–135
< 1.2	< 2.4	ID	ID	> 0.60	> 31	20–40	55–135
1.4 ± ID	ID	39 ± 4	ID	0.79 ± 0.02	~ 38	20–40	55–110
0.8 ± 0.3	ID	23 ± 6	ID	0.74 ± 0.02	~ 38	10–40	40–110
1.2 ± 0.1	3.0 ± 0.4	51 ± 6	20 ± 7	0.67 ± 0.02	~ 34	10–40	25–110
1.2 ± 0.2	2.4 ± 0.5	75 ± 15	11 ± 6	0.60 ± 0.07	~ 31	5–20	25–80
1.5 ± 0.2	2.6 ± 0.3	67 ± 10	9 ± ID	0.62 ± 0.04	~ 32	15 or less	
1.4 ± 0.2	2.6 ± 0.4	87 ± 10	13 ± 2	0.54 ± 0.04	28	15 or less	15–40
ID	ID	ID	ID	ID	—	5 or less	15–25
2.0 ± 1.2	3.8 ± 0.8	72 ± 30	20 ± 9	0.47 ± 0.05	~ 25	10 or less	15–25
2.6 ± 1.3	3.9 ± 1.5	103 ± 34	11 ± 6	0.35 ± 0.09	~ 19	15 or less	15–40
ID	ID	ID	ID	ID	—	5 or less	5–25
—	—	—	—	—		—	—

Table 2.2—*continued*

(d) Engineering use chart

Typical names of soil groups	Group symbols	Permeability when compacted	Shear strength when compacted and saturated	Compressibility when compacted and saturated	Workability as a construction material
Well-graded gravels, gravel–sand mixtures, little or no fines	GW	Pervious	Excellent	Negligible	Excellent
Poorly-graded gravels, gravel–sand mixtures, little or no fines	GP	Very pervious	Good	Negligible	Good
Silty gravels, poorly-graded gravel–sand–silt mixtures	GM	Semipervious to impervious	Good	Negligible	Good
Clayey gravels, poorly-graded gravel–sand–clay mixtures	GC	Impervious	Good to fair	Very low	Good
Well-graded sands, gravelly sands, little or no fines	SW	Pervious	Excellent	Negligible	Excellent
Poorly-graded sands, gravelly sands, little or no fines	SP	Pervious	Good	Very low	Fair
Silty sands, poorly-graded sand–silt mixtures	SM	Semipervious to impervious	Good	Low	Fair
Clayey sands, poorly-graded sand–clay mixtures	SC	Impervious	Good to fair	Low	Good
Inorganic silts and very fine sands, rock flour, silty or clayey fine sands with slight plasticity	ML	Semipervious to impervious	Fair	Medium	Fair
Inorganic clays of low to medium plasticity, gravelly clays, sandy clays, silty clays, lean clays	CL	Impervious	Fair	Medium	Good to fair
Organic silts and organic silt–clays of low plasticity	OL	Semipervious to impervious	Poor	Medium	Fair
Inorganic silts, micaceous or diatomaceous fine sandy or silty soils, elastic silts	MH	Semipervious to impervious	Fair to poor	High	Poor
Inorganic clays of high plasticity, fat clays	CH	Impervious	Poor	High	Poor
Organic clays of medium to high plasticity	OH	Impervious	Poor	High	Poor
Peat and other highly organic soils	Pt	—	—	—	—

Important engineering properties

Source: United States Bureau of Reclamation (1974).

Relative desirability for various uses
(No. 1 is considered the best)

Rolled earthfill dams			Canal sections		Foundations		Roadways		
								Fills	
Homogeneous embankment	Core	Shell	Erosion resistance	Compacted earth lining	Seepage important	Seepage not important	Frost heave not possible	Frost heave possible	Surfacing
—	—	1	1	—	—	1	1	1	3
—	—	2	2	—	—	3	3	3	—
2	4	—	4	4	1	4	4	9	5
1	1	—	3	1	2	6	5	5	1
—	—	3 if gravelly	6	—	—	2	2	2	4
—	—	4 if gravelly	7 if gravelly	—	—	3	6	4	—
4	5	—	8 if gravelly	5 erosion critical	3	7	8	10	6
3	2	—	5	2	4	8	7	6	2
6	6	—	—	6 erosion critical	6	9	10	11	—
5	3	—	9	3	5	10	9	7	7
8	8	—	—	7 erosion critical	7	11	11	12	—
9	9	—	—	—	8	12	12	13	—
7	7	—	10	8 volume change critical	9	13	13	8	—
10	10	—	—	—	10	14	14	14	—
—	—	—	—	—	—	—	—	—	—

Table 2.2—continued
(e) Typical correlations

Group symbol†	Typical values						Other classifications	
	LL	PI	LS	Activity	C_c	c_v (cm^2 s^{-1})	Revised PRA class	SA railway‡
GW	NP	—	—					A3	A
GP	NP	—	—					A1	A
GM	NP	—	—					A2	A
GC	<30	<15	<8					A3	A
SW	NP	—	—					A3	A
SP	NP	—	—					A1	A
SM	10–30	<4	—					A3	A
SC	10–20	7–12	<6	<1.0	0.20	10^{-2}	A2	A
ML	20–50	<20	<6	0.25–0.75	0.20	10^{-2}	A4, A6, A7	B
CL	20–50	6–25	<10	0.25–1.0	0.30	10^{-3}	A4, A6, A7	B
OL	20–50	<20	<6	0.25–0.75	0.35	10^{-3}	A4, A7	B
MH	>50	<30	<12	0.5–1.0	0.40	10^{-3}	A5	C
CH	50–100+	20–50+	8–15+	0.5–2.0+	0.50	10^{-4}	A6, A7	C
OH	>50	<30	<12	0.5–2.0	1.0	10^{-4}	A7	C
Pt	100–800	60–120?	?	—	2–4	—	A7	C

† Source: United States Army Corps of Engineers (1953).
‡ Source: Rauch (1963).
NP = non-plastic.

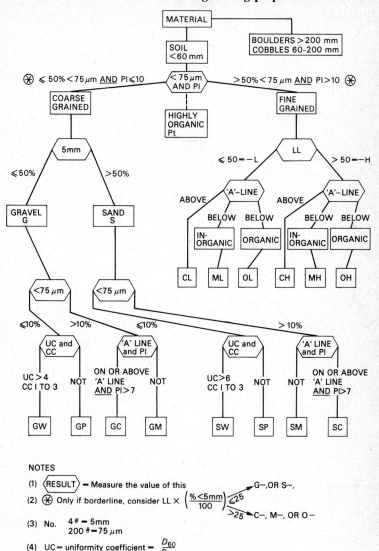

NOTES

(1) ⟨RESULT⟩ — Measure the value of this

(2) ✳ Only if borderline, consider $LL \times \left(\dfrac{\% <5mm}{100} \right)$ $\begin{cases} \leqslant 25 \\ >25 \end{cases}$ → G–, OR S–, → C–, M–, OR O–

(3) No. $4\# = 5mm$
$200\# = 75 \mu m$

(4) UC = uniformity coefficient = $\dfrac{D_{60}}{D_{10}}$

CC = coefficient of curvature = $\dfrac{(D_{30})^2}{D_{60} \times D_{10}}$

(5) D_{60} means that 60 per cent of the particles are finer this size.

(6) The A–line is defined on the plasticity chart in Table 2.2

FIG. 2.13. Flow chart for rapid determination of soil class within the unified classification system. (Our metrication.)

the material within a particular group. It is a way of classifying the quality of a road subgrade material within its own group and the higher the value (maximum is 20) the poorer is the quality; the value is determined from Fig. 2.14. A quick guide to the identification of groups in this system is given in Fig. 2.15. In Fig. 2.1 the clay would belong to the A-7-5(20), the sand to the A-2-4(0) and the gravel to the A-1-a class.

RECOMMENDED READING

A general textbook, recently revised, for student, practioner, or specialist is:

Terzaghi, K. and Peck, R. B. (1967). *Soil mechanics in engineering practice*, 2nd edn. Wiley, New York.

A useful review of the present state of the art in regard to bearing capacity and settlement analysis is contained in:

Simon, N. E. and Menzies, B. K. (1975). *A short course in foundation engineering*. IPC Science and Technology Press, Guildford, Surrey.

Other works include:

Akroyd, T. N. W. (1957). *Laboratory testing in soil engineering*. Foulis, London.
Winterkorn, H. F. and Fang, H-Y. (eds.) (1975). *Foundation engineering handbook*. Van Nostrand Reinhold, New York

On certain aspects progress is still being made so rapidly that up-to-date information is contained only in the Proceedings of the four-yearly International Conferences on Soil Mechanics and Foundation Engineering or the intervening Regional Conferences.

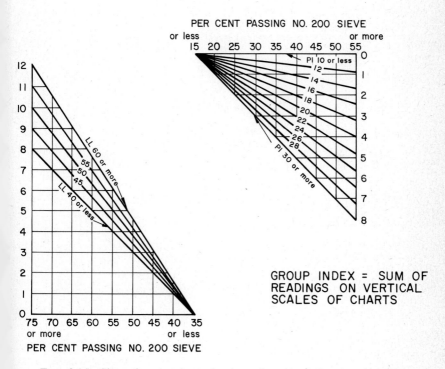

FIG. 2.14. Chart for the determination of group index for the PRA classification system.

TABLE 2.3

The PRA (Public Roads Administration) Classification is a useful guide to the use of soils as road construction materials. (Modified from Allen 1945)

Group	Sub-group	Per cent passing US sieve number			Character of fraction passing number 40 sieve		Group index number	Soil description	Subgrade rating
		10	40	200	Liquid limit	Plasticity index			
A-1		50 max	50 max	25 max		6 max	0	Well-graded gravel or sand; may include fines	
	A-1-a		50 max	15 max		6 max	0	Largely gravel but can include sand and fines	
	A-1-b		50 max	25 max		6 max	0	Gravelly sand or graded sand; may include fines	
A-2†				35 max			0 to 4	Sands and gravels with excessive fines	Excellent to good
	A-2-4			35 max	40 max	10 max	0	Sands, gravels with low-plasticity silt fines	
	A-2-5			35 max	41 min	10 max	0	Sands, gravels with elastic silt fines	
	A-2-6			35 max	40 max	11 min	4 max	Sands, gravels with clay fines	
	A-2-7			35 max	41 min	11 min	4 max	Sands, gravels with highly plastic clay fines	
A-3			51 max	10 max		Nonplastic	0	Fine sands	

A-4	36 min	40 max	10 max	8 max	Low-compressibility silts	Fair to poor
A-5	36 min	41 min	10 max	12 max	High-compressibility silts, micaceous silts	
A-6	36 min	40 max	11 min	16 max	Low-to-medium compressibility clays	
A-7	36 min	41 min	11 min‡	20 max	High-compressibility clays	
A-7-5	36 min	41 min	11 min‡	20 max	High-compressibility silty clays	
A-7-6	36 min	41 min	11 min	20 max	High-compressibility, high-volume-change clays	
A-8					Peat, highly organic soils	Unsatisfactory

† Group A-2 includes all soils having 35 per cent or less passing a number 200 sieve that cannot be classed as A-1 or A-3.
‡ Plasticity index of A-7-5 subgroup is equal to or less than LL-30. Plasticity index of A-7-6 subgroup is greater than LL-30.

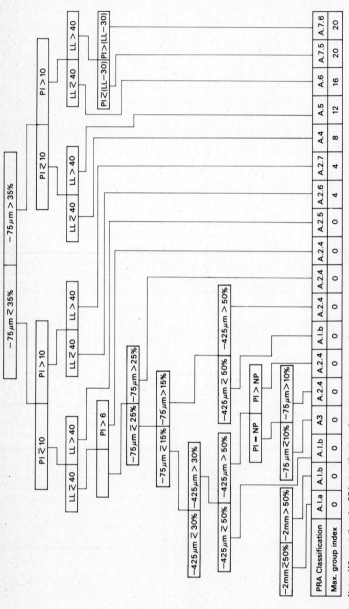

Note: When stating the PRA classification of a subgrade the group index is also stated in parenthesis after the PRA group number. (Prepared by Van Niekerk, Kleyn and Edwards)

Fig. 2.15. Flow chart for the rapid determination of soil class within the PRA classification system. The minus sign (as in −75 μm ≲35%) means 'finer than'.

3. Introduction to rocks, soils, and landforms

The basic unit of foundation material, and often also of potential construction material, with which the engineer is concerned at any site is the *soil profile*, which comprises all horizons of unconsolidated material down to and including the surface of the underlying bedrock.

In order fully to appreciate the nature of the soil profile at any specific site it is necessary to understand the processes which have given rise to the different horizons of soil and rock comprising it as well as their genetic relationships.

The genetic cycle responsible for the formation of different types of rocks and soils is illustrated in the flow diagram in Fig. 3.1 and in the schematic cross-sections of Fig. 3.2. The geological processes involved in this cycle that are of particular significance to the soil surveyor and the engineer are discussed below, but first we give a brief account of all the processes and their products.

The cycle begins with the magma, a viscous melt of one silicate in another, deep beneath the earth's continental crust. Crystallization of minerals from a magma produces *igneous rock*. In the zone of weathering near the surface, rock weathers by mechanical disintegration and chemical decomposition to form residual soil. *Residual soil* is thus decomposed rock matter *in situ*. It consists partly of unaltered primary minerals which were present in the parent rock (e.g. quartz) and partly of secondary minerals (mainly clay minerals) which have formed by hydration, hydrolysis, carbonation, oxidation, and reduction, i.e. reactions with water, carbon dioxide, and oxygen.

Residual soil removed from its place of origin by natural agencies is deposited elsewhere as unconsolidated sediment. In engineering terminology, *all* unconsolidated sediments, whether deposited as shallow terrestrial deposits (e.g. pediment deposits) or as deep accumulations of marine sediments, are classified as *transported soils*. The term *soil* is thus used by engineers in a very different way from other earth scientists, and includes all unlithified or unconsolidated materials, whether residual or transported.

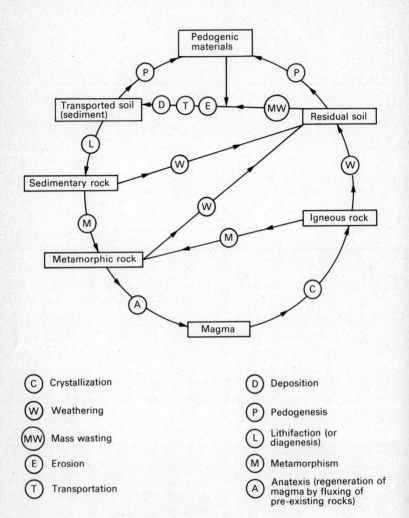

FIG. 3.1. The 'rock–soil' genetic cycle.

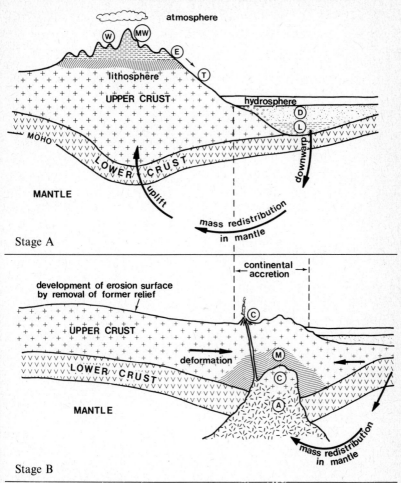

FIG. 3.2. The genesis of different rock types through the action of various geological processes. A hiatus of hundreds of millions of years separates stage A from stage B.

The local, short distance, agent of removal of rock and residual soil is gravity, and the processes, e.g. landslides, or soil creep, often aided by sheetwash, are collectively known as *mass wasting* processes. The long distance transportation agents are water, wind, and ice. Materials eroded, transported, and deposited elsewhere as transported soils (or sediments) may subsequently become lithified into *sedimentary rocks*—for example by the deposition of a calcareous or siliceous cementing agent in the voids between the discrete particles, or by extreme overconsolidation under overburden pressure.

Sedimentary rock or igneous rock which is subjected to a sufficiently high degree of stress and/or temperature will alter into *metamorphic rock*, and, if the temperature under the prevailing pressure rises above the melting point of the minerals concerned, e.g. as a result of the deep downwarp of a marine sedimentary basin (Fig. 3.2), a magma will once more be generated by the process of anatexis.

Finally, *pedogenesis*—resulting mainly from the leaching action of water percolating through either residual or transported soil—may form pedogenic materials within the upper horizons of the soil profile (Fig. 3.1). Certain hard varieties of pedogenic materials, such as calcrete, silcrete, ferricrete or laterite, are referred to as *pedocretes*.

COMPOSITION OF ROCKS

Rocks are aggregates of mineral particles. A *mineral* is an inorganic substance with a fixed chemical composition and consistent physical properties: it is a chemical compound composed of *elements* in fixed proportions by mass.

The eight elements listed in Table 3.1 comprise 98.6 per cent by mass of the rocks of the earth's continental crust. The twelve most common minerals, which together make up 99 per cent of all rocks (and soils) in the earth's crust (Table 3.2), are composed of two or more of the eight elements listed in Table 3.1, in some cases together with hydrogen and carbon. From the relative abundance of these elements it is not surprising that the most common rock-forming minerals are quartz (SiO_2) and silicates and alumino-silicates of iron, calcium, sodium, potassium, and magnesium.

TABLE 3.1

Common elements of the earth's crust

Elements in the earth's crust		Mass (%)
O	oxygen	46.7
Si	silicon	27.7
Al	aluminium	8.1
Fe	iron	5.0
Ca	calcium	3.6
Na	sodium	2.8
K	potassium	2.6
Mg	magnesium	2.1
		98.6

Table 3.2 presents some physical properties of the minerals which may be an aid to the identification of hand specimens.

We return now to the starting point of the genetic cycle given in Fig. 3.1.

CRYSTALLIZATION: IGNEOUS ROCKS

The order in which the common minerals of igneous rocks crystallize from a cooling magma is indicated in Table 3.3. Olivine and calcic plagioclase are the first minerals to crystallize in abundance from a magma with a basic (or basaltic) composition. Unless these crystals settle to the floor of the magma chamber, the olivine reacts with the remaining magma as it cools further to produce pyroxene, and the calcic plagioclase reacts to produce a feldspar richer in sodium. If no crystals settle from the melt, these reactions go on unimpeded to produce a rock consisting of a mixture of plagioclase and pyroxene with a small amount of olivine, e.g. basalt. But if the crystals which formed early separate from the magma in large quantities by settling out, such reactions cannot take place, and the remaining magma may finally solidify to form a rock of an intermediate composition (e.g. andesite) or even of an acidic composition (e.g. granite). It is usually only the acid igneous rocks which contain free silica in the form of quartz, as this is the last mineral to crystallize from the magma.

The order of crystallization has a profound effect on the

TABLE 3.2

Composition and properties of the common rock-forming minerals

	Mineral	Composition	Mohs' hardness	SG	Fracture
	Microcline and orthoclase (feldspar)	$KAlSi_3O_8$	6	2.6	Good cleavage at right angles
	Plagioclase (feldspar)	$NaAlSi_3O_8$ $CaAl_2Si_2O_8$	6	2.7	Cleavage nearly at right angles
	Quartz	SiO_2	7	2.65	Conchoidal fracture
Minerals forming igneous rocks	Muscovite (white mica)	Basic silicate of K, Al	2.5	2.8	Perfect single cleavage
	Biotite (black mica)	Basic silicate of Mg, Al, Fe	3	3.0	Perfect single cleavage
	Hornblende (amphibole)	Silicate of Ca, Mg, Fe, Al	5-6	3.05	Good cleavage at 120°
	Augite (pyroxene)	Silicate of Ca, Mg, Fe, Al	5-6	3.05	Good cleavage nearly at 90°
	Olivine	$(Mg, Fe)_2SiO_4$	6-7	3.5	Conchoidal fracture
Additional minerals common in some sedimentary rocks	Calcite	$CaCO_3$	3	2.7	Three perfect rhomboidal cleavages
	Dolomite	$CaCO_3 . MgCO_3$	4	2.8	Three perfect rhomboidal cleavages
	Kaolinite	$(OH)_8Al_4Si_4O_{10}$	1	2.6	
	Haematite	Fe_2O_3	6	5.0	No cleavage

susceptibility of the different minerals to decomposition in the zone of weathering.

An elementary classification of common igneous rocks is given in Table 3.4.

TABLE 3.3
Order of crystallization of minerals from a magma

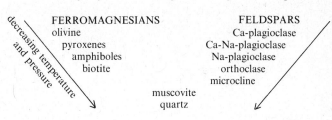

	FERROMAGNESIANS		FELDSPARS
decreasing temperature and pressure	olivine		Ca-plagioclase
	pyroxenes		Ca-Na-plagioclase
	amphiboles		Na-plagioclase
	biotite		orthoclase
			microcline
		muscovite	
		quartz	

WEATHERING OF ROCK

Weathering processes may be classified as physical or chemical depending on whether the parent rock undergoes alteration by mechanical disintegration or by chemical decomposition. Of these, chemical decomposition is by far the more important in the creation of residual soil, though the division between the two classes of processes is somewhat arbitrary since mechanical disintegration is responsible for an increase in the specific surface area available for chemical attack. Furthermore, most chemical processes produce secondary minerals with larger volumes than their parent minerals, thus creating stresses which aid mechanical disintegration. In short, physical and chemical weathering processes often taken place simultaneously and they aid and abet one another.

Five factors control the nature and rate of weathering processes. The nature of the *parent material*, i.e. rock type and rock structure, is initially the most important factor. Mineralogical composition determines, in part, whether the rock is more susceptible to mechanical disintegration or to chemical decomposition, while discontinuities in the rock mass, such as joints and bedding planes, facilitate the entry of atmospheric reagents into the rock. *Climate* not only governs the type of weathering which is likely to take place

TABLE 3.4
Elementary classification of igneous rocks

		ALKALINE / ACID	INTERMEDIATE	BASIC	ULTRABASIC
PREDOMINANT DARK MINERALS		Biotite, hornblende, present only in small amounts	Hornblende, biotite, pyroxene	Usually pyroxene. Hornblende and biotite rare.	Pyroxene, olivine, some ore minerals (magnetite, chromite etc.)
COMPOSITION OF FELDSPAR		Orthoclase/microcline, sodic plagioclase (albite-oligoclase)	Intermediate plagioclase (andesine)	Calcic plagioclase (labradorite)	Little or no feldspar
OVERSATURATED (contains quartz)	Plutonic (grains >2 mm)	Granite	Granodiorite		
	Hypabyssal (grains 0.1–2 mm)	Aplite			
	Volcanic (grains <0.1 mm)	Rhyolite			
SATURATED (contains no quartz, olivine or nepheline)	Plutonic	Syenite	Diorite	Gabbro	Pyroxenite
	Hypabyssal	Syenite–aplite	Diabase	Dolerite	
	Volcanic	Trachyte	Andesite	Basalt	
UNSATURATED (contains olivine or nepheline)	Plutonic	Nepheline-syenite			Peridotite
	Hypabyssal	Tinguaite			Kimberlite
	Volcanic	Phonolite			Kimberlite

Na and K decreasing; Mg and Fe increasing →

but also influences its rate: warm, humid conditions favour chemical decomposition, while arid conditions favour mechanical disintegration.

Living organisms, particularly the vegetative cover, may modify the climatic influence by affecting runoff and infiltration of rainfall. The wedging action of roots in rock fissures contributes to mechanical disintegration; organic matter provides a source of carbon dioxide and a variety of organic acids which are active in chemical decomposition. *Topography* may either accelerate or retard weathering processes: in flat terrain, rain water is lost much more slowly than in a rolling landscape. On the other hand, if a given area is perennially waterlogged, chemical decomposition of the underlying rock is retarded. And, finally, the period of *time* over which the 'active' weathering factors of climate and living organisms have acted upon the 'passive' factors of parent material and topography will determine the degree of weathering. As discussed later, the effect of the time factor is often the crucial difference between landscapes of different geomorphological age.

Mechanical disintegration

Physical weathering, or mechanical disintegration, gives rise to *in situ* fragmentation of rock without significant mineralogical or chemical changes. These processes include:

Unloading due to removal of overburden by erosion

Expansion of the rock mass on the relief of overburden pressure by the erosion of overlying rock is perhaps the most important of these processes. Most characteristic in plutonic masses of granite–gneiss, this process results in the formation of joint sets roughly parallel to the surface and becoming more widely spaced with depth. This commonly produces large-scale exfoliation in sheet form, which may extend to a substantial depth (>500 m) below the surface. In sedimentary rocks, relief of overburden pressure may result in concentric jointing coupled with granular disintegration.

Mechanical effects of chemical decomposition

Volumetric increases attendant upon the chemical process of hydration result in stresses within the mass of weathering rock, the development of weathering spheroids within basic igneous rocks, and the spalling-off of the outer portions of the rock.

Growth of ice crystals and salt crystals

Externally induced stresses also contribute to physical weathering. The formation of ice crystals in fissures in rock increases the volume of the water by 9.05 per cent, which may produce pressures exceeding the tensile strength of the rock. Unless the freezing mass is completely confined, however, the resulting pressures may be absorbed by expulsion of water or compression of air. Similar processes are encountered in desert environments where salts and other evaporites are concentrated in joints and fractures through capillary action.

Biotic activity

The influence of plant roots in widening cracks and crevices in rock is most evident in the shattering action of the roots of trees growing in thin soils over rock, but it is likely that similar disruptive forces are exerted on a smaller scale by any plant roots that penetrate the interstices of rocks. The action of worms, termites and burrowing animals may also have mechanical effects, and such activity serves chemical decomposition most effectively by exposing fresh surfaces to chemical attack.

Thermal expansion and contraction

It seems likely that diurnal temperature fluctuations cannot produce stresses large enough to overcome the elastic properties of most rocks. However, in tropical environments a sudden downpour of rain onto a solar-heated rock surface may sometimes produce spalling.

Chemical decomposition

Minerals in the zone of weathering react with water, oxygen, and carbon dioxide at atmospheric temperatures and pressures to produce residual soils.

Minerals which crystallized during the early stages of magma cooling are more susceptible to these chemical reactions than are those that crystallized late (Table 3.3). This is because the reactions take place at atmospheric temperature and pressure—conditions more nearly approaching those that obtained during the final stages of magma crystallization. Thus olivine and calcic plagioclase decompose far more rapidly than quartz.†

The primary minerals in the zone of weathering decompose as a result of the hydration of cations, oxidation (which reduces the size of cations), and hydrolysis which replaces cations by hydrated hydrogen ions or breaks the links between cations. The residual soil produced is a mixture of resistant primary minerals such as quartz, insoluble weathering products such as alumina or silica, and new or secondary minerals such as clays, and contains soluble products such as the chlorides, sulphates and bicarbonates of sodium, potassium, magnesium, or calcium which may subsequently be leached out.

For example, feldspars and ferromagnesian minerals break down to form clay minerals, and the further transformation of these into other clay minerals involves hydrolysis followed by a progressive removal of bases.

$$4KAlSi_3O_8 + 6H_2O = (OH)_8Al_4Si_4O_{10} + 8SiO_2 + 4KOH$$

orthoclase $\qquad$ kaolinite $\qquad$ colloidal $\quad$ solute
$\qquad\qquad\qquad\qquad\qquad\qquad\qquad\qquad$ silica

$$CaAl_2Si_2O_8 . 2NaAlSi_3O_8 + nH_2O = (OH)_4Al_4Si_8O_{10} . nH_2O$$

plagioclase $\qquad\qquad\qquad\qquad\qquad\qquad$ montmorillonite

$$+ Ca(OH)_2 + 2NaOH$$

solutes

In the initial stages the primary minerals may break down to 2:1 lattice clays (chlorite, vermiculite, illite, montmorillonite) in which one layer of alumina octahedra is sandwiched between two layers of silica tetrahedra in an arrangement which requires fewer balancing cations than the parent silicates. Further removal of bases results in the formation of the 1:1 lattice of kaolinite, in which each lattice sheet consists of one layer of silica tetrahedra attached to one layer of alumina octahedra with a charging balance much lower than in the 2:1 lattice clays. Thus far fewer balancing cations have to be incorporated in the lattice. Prolonged leaching in a tropical environment ultimately leads to the breakdown of kaolinite by the removal of silica to leave a residue of hydrated alumina, i.e. 'hardpan' gibbsite.

---

† It is for this reason that river and beach sands usually consist for the most part of the mineral quartz; where feldspar is present in such deposits it is most likely to be microcline.

Susceptibility of different rock types to weathering

Since the secondary minerals formed by the processes of chemical weathering are chemically more stable under atmospheric conditions than the primary minerals (with the exception of quartz), it follows that sedimentary rocks, which mostly consist of secondary minerals will, on the whole, be less susceptible to chemical weathering than igneous rocks and certain metamorphic rocks. This does not apply to calcareous sedimentary rocks such as chalk, limestone or dolomite, which are soluble in water containing carbon dioxide in solution. On the other hand, most sedimentary rocks contain planes of structural weakness such as bedding planes and lack interlocking mineral particles, which makes them more vulnerable to the processes of mechanical disintegration than igneous rocks and massive metamorphic rocks. Thus it is possible to make broad generalizations about the relative susceptibility of different rock types to weathering in different climates, as in Table 3.5, and the relative depths of residual soil mantles likely to be encountered are also implicit in this table.

TABLE 3.5

Relative susceptibility of different rock types to weathering in desert and tropical environments. 1: most vulnerable, 5: least vulnerable

		Susceptibility to weathering in	
Rock category	Common examples	Desert environment (mechanical disintegration dominant)	Tropical environment (chemical decomposition dominant)
Acid crystalline rock	Granite; gneiss	3	3
Basic igneous rock	Basalt; dolerite	4	2
Calcareous rock	Chalk; limestone; dolomite; marble	5	1
Argillaceous sedimentary rock	Mudstone; shale	1	4
Arenaceous sedimentary or metamorphic rock	Quartzite; sandstone	2	5

RESIDUAL SOILS

Residual soil is thoroughly decomposed rock *in situ*, and this generally grades downwards through zones of progressively less weathered rock and finally into fresh bedrock. In engineering practice, residual soils are generally named after the parent rock from which they were derived. Table 3.6 lists the soil textures of some common examples. Indeed it is often possible to predict the

TABLE 3.6
Types of residual soil with corresponding soil textures

Parent rock category	Common examples of residual soils	Soil texture
Acid crystalline rock	Residual granite; residual gneiss	Clayey sand or sandy clay (often micaceous)
Basic igneous rock	Residual gabbro; residual dolerite; residual basalt	Clay (often grading into sandy clay with depth)
Calcareous rock	Residual limestone; residual dolomite	Gravel in clayey or silty matrix
Argillaceous sedimentary rock	Residual mudstone; residual shale	Silt or silty clay
Arenaceous sedimentary or metamorphic rock	Residual sandstone; residual quartzite	Sand (clayey sand in the case of residual arkose or feldspathic sandstone)

texture of a residual soil and its likely engineering properties from the mineralogical composition of its parent rock, a principle which may be applied to advantage in the interpretation of aerial photographs. For example residual granite usually consists of a mixture of kaolinite and quartz in the form of a clayey sand, whereas basic igneous rocks like basalt or dolerite, which contain no quartz, decompose to form clay, often of the active or expanding lattice type.

TRANSPORTED SOILS

Residual soil which has been removed from its place of origin by the processes of mass wasting and erosion is deposited elsewhere as transported soil (or sediment). The most common types of transported soil deposits, the agencies responsible for their deposition and their probable textures are listed in Table 3.7 more or less in

TABLE 3.7
Origins of transported soils

Transported soil type	Agency of transportation	Source of parent material	Probable soil texture
Biotically reworked soil	Termites, worms, etc.	Underlying soil	Usually clayey or silty sand
Slide debris	Mass wasting processes	Soil and/or rock debris immediately upslope	Mixture of soil and poorly sorted rock debris
Talus (coarse colluvium)	Gravity (mass wasting processes)	Rock or weathered rock outcropping immediately above the talus slope	Unsorted angular gravel and boulders usually with a finer textured matrix
Pediment deposit (fine colluvium)	Sheetwash	Talus deposit immediately above pediment	Predictable from knowledge of source rock: e.g. clayey sand from granite, clay from dolerite, silt from shale (see textural classes in Table 3.6)
Pyroclastic deposit	Volcanism	Volcanic ash and other ejectamenta	Silt and sand and cinder gravels sometimes containing larger fragments
Gully-wash	Flow in steep-gradient gullies	Soil in local gully-catchment	Unsorted gravels, sands, silts and clays

Till	Glacial ice	All soil and rock types along the path of glacial advance	Mixture of rock and soil debris of great textural diversity
Aeolian deposit	Wind	Usually mixed source; moraines and glacial outwash in the case of loess	Dune deposits of sand; loose deposits of silt and fine sand containing large quantities of primary minerals where deposit is not weathered
Alluvium	Streams	All materials within the catchment	Usually gravels and sands in fan deposits. Gravels and sands in terraces. Sand in levees. Silts or clays in back-swamp and flood-plain deposits
Lacustrine deposit	Streams depositing in lakes, pans, 'vleis' and 'dambos'	All materials within the catchment	Sands, silts, clays
Estuarine and deltaic deposits	Tidal rivers depositing into saline water	All materials within the catchment	Usually clays and silts deposited by flocculation in brine. Sandy facies (finger sands and pro-deltaic beds) sometimes present
Littoral and marine deposits	Waves, currents and tides	Materials carried to the sea rivers; diverse source	Beach sand. Storm beach gravels. Marine deposits

order of their distance of transportation. Unless the regime of the transporting agency precludes sorting—e.g. episodic rivers will tend to deposit poorly sorted alluvium—the greater the distance of transportation the greater will be the degree of sorting into uniform grain sizes, except that glacial till, though transported in places for hundreds of kilometres, remains the most unsorted conglomeration of sizes and mineralogical composition of all transported materials.

The thicknesses of transported soil deposits vary greatly but gully-wash and pediment deposits are generally shallow (usually no more than a few metres), slide debris, talus, and alluvium are commonly of intermediate thickness, and the other deposits listed in Table 3.7 are often deep (tens of metres or, in the case of marine deposits, hundreds of metres).

Chapter 5 shows that specific engineering problems are commonly associated with particular types of transported soils. So a correct identification of the *origin* of a transported soil, and indeed also of a residual soil in a given environment, enables the engineer to predict likely engineering problems. The recognition of the main elements of the local physiography is often a great aid in the identification of transported soil type. This is illustrated diagrammatically in Fig. 3.3.

In soil profiles in a wide variety of environments a band of gravel is found at the junction of the transported soil and the residual soil. In many parts of Africa this feature has become known as the *pebble marker* (Fig. 3.4). It represents the most recent major geological unconformity in the soil profile, occurs at the base of the Quaternary (transported) deposits and is thus less than about two million years old, though in places it is still forming today (Brink 1979). Though it varies in thickness and mode of origin from one place to another, the identification of the pebble marker within the soil profile is of major significance in that it invariably represents the horizontal boundary between soils with substantially different engineering characteristics. In one area it may consist of well-rounded gravels representing the base of an alluvial terrace, e.g. the Thames gravels between the underlying Oxford Clay of Jurassic age (200 Ma) and the overlying solifluction debris of Pleistocene age (< 1 Ma). In another it may be a colluvial deposit of angular gravel transported down the pediment under the influence of gravity and rainwash. Below ancient erosion surfaces in tropical and semi-tropical environments the pebble marker is often represented by a

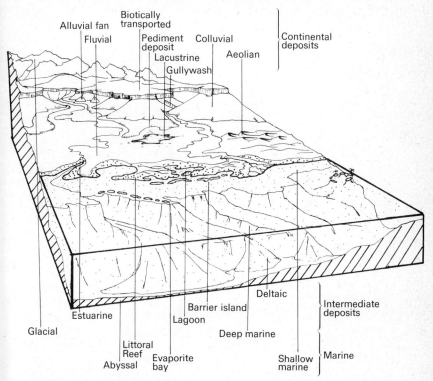

FIG. 3.3. Sedimentary environments and landforms associated with various transported soils.

biogenic stone line formed by the action of termites, which have carried the finer particles of soil upwards to form a horizon of biotically transported soil and to leave a concentration of stones at the lower limit of their zone of activity. However, only where the biogenic stone line is underlain by residual soil does it constitute the pebble marker. In situations where the residual soil is a stone-free material the pebble marker may be represented by isolated pebbles or shells, or even artefacts, scattered upon the pre-Quaternary erosion surface. Considerable care may be necessary to distinguish the pebble marker, which is that particular gravel layer sandwiched between the transported and residual soils, from other gravelly horizons within the transported zone and sometimes even in the residual zone of a soil profile.

FIG. 3.4. A pebble marker of vein quartz gravel separates fine sandy colluvium in the upper half of the profile from clayey soil residual from Precambrian andesite in the lower half. (Photograph by A. A. B. Williams).

PEDOGENESIS

Pedogenesis re-arranges materials within the soil profile, particularly by the leaching of components by rainwater percolating through the soil. Soluble substances are the first to be leached out, but soil organic matter may produce chelating substances, or organic acids, with specific abilities for mobilizing iron, manganese, or aluminium, particularly in intermittently and permanently waterlogged situations, and colloidal materials are then translocated downwards within the soil profile. Thus the upper layers of a soil profile, in either transported soil or residual soil, may be re-arranged into a surface horizon of *eluviation* which has lost materials to the subsoil horizon of *illuviation*.

Living organisms such as worms, termites, and rodents also rearrange soil particles, which increases the void ratio of the soil, contributes to its aeration and promotes the other pedogenic processes associated with leaching. In extreme cases such bioturbation may cause large scale vertical movement of soil to create a biotically transported horizon with a basal biogenic stone line.

Pedogenesis is of particular significance in engineering when it results in the formation of pedocretes, i.e. nodular or 'hardpan' layers of calcrete, silcrete, ferricrete, or laterite. Pedocretes are highly valued construction materials in many parts of the world, especially for road construction, and they often provide competent founding materials for a variety of engineering structures.

LITHIFACTION: SEDIMENTARY ROCKS

As illustrated in Fig. 3.3, transported materials may be deposited in continental, intermediate, or marine environments. In favourable localities these deposits build up in layers and become lithified to form sedimentary rocks. Lithifaction (or diagenesis) may be achieved by physical compaction under the load of overlying sediment, or by cementation involving the deposition of mineral matter in the voids between the sedimentary particles. Consolidation operates in thick accumulations of clays and silts as these materials are relatively impervious, though initially highly porous. Sand and gravel deposits, on the other hand, are usually well packed and undergo little consolidation; however, being highly permeable materials, they are particularly susceptible to cementation.

Marine environments are the most favourable for the accumulation of thick deposits of transported material and thus produce the vast bulk of sedimentary rocks (Fig. 3.3). Littoral or beach deposits accumulate in the intertidal zone. Shallow marine (shoalwater) deposits accumulate on the continental shelf which generally extends to a depth of about 200 m below sea level; sands and shell fragments are laid down relatively close to the shore, muds are deposited further out, and calcium carbonate, which ultimately forms limestones, is precipitated yet further out. Abyssal deposits are laid down in the ocean depths; they include siliceous and calcareous oozes composed of the remains of minute marine organisms, and muds formed from volcanic dust settling in the deep waters.

An elementary classification of sedimentary rocks is given in Table 3.8, and of metamorphic rocks in Table 3.9.

LANDFORMS AND EROSION SURFACES

The distribution and properties of soils can only be comprehended in the context of the landscape in which they develop. In sub-humid tropical environments, for instance, on the well-drained upper slopes or crestal areas which have not been subjected to erosion, leaching tends to produce granular, well-drained soils with a high proportion of voids. Mid-slope zones may be subject either to erosion or deposition and the degree of leaching may be expected to be less advanced; seasonal groundwater effects may produce well-developed pedogenic horizons in the upper layers of the soil profile. In poorly drained areas and bottomlands groundwater effects are more prominent; the introduction of clay minerals and the retardation of leaching processes is likely to produce active clay minerals such as montmorillonite. This idealized sequence, or *catena*, from uplands to bottomlands is commonly parallelled by changes in soil colour from reds through yellows to greys or whites. Usually each zone supports a characteristic plant association and presents a specific range of engineering properties, e.g. deep compressible soils often with a collapsible fabric (Chapter 5) are likely to be associated with deeply leached upper slopes, and potentially expansive soils with the bottomland areas. Hard ferricrete for use as road construction materials would be sought on the middle and lower slopes (Fig. 3.5).

TABLE 3.8

Elementary classification of sedimentary rocks. (Modified from Miller and Scholten 1962)

Texture			Composition of clastic rocks				
		Grains	mainly quartz	quartz and microcline	quartz, chert and rock fragments; sometimes plagioclase	calcite and/or dolomite	modified plant fragments
		Matrix	silty, clayey, and/or ferruginous, or absent	clayey, silty, ferruginous	clayey, silty, carbonaceous	clayey, silty, carbonaceous	
		Cement	calcareous, siliceous, and/or ferruginous	calcareous or absent	commonly absent	calcareous; rarely siliceous	
clastic		mainly gravel-size particles (>2.0 mm)	Quartz conglomerate (breccia if angular)	Arkose conglomerate (breccia if angular)	Greywacke conglomerate (breccia if angular) Tillite	Limestone or dolomite conglomerate or breccia (var. shelly limestone)	
		mainly sand-size particles (2.0 mm to 0.6 mm)	Sandstone (>85% quartz and chert) Orthoquartzite (>95% quartz and chert) Calcarenite (calcareous cement)	Felspathic sandstone (<30% feldspar) Arkose (>30% feldspar)	Greywacke	Medium-clastic limestone or dolomite	Peat, lignite and coal
		mainly silt-size particles (0.06 mm to 2 μm)	Siltstone			Limestone (var. chalk)	
		mainly clay-size particles (<2 μm)	Mudrock (shale with bedding planes; mudstone without)			Dolomite	
crystalline		fine (<0.06 mm)	Chert	Sedimentary hematite	Rock salt (halite)	Crystalline limestone (var. travertine)	
		medium (0.06–2.0 mm)		Bog iron ore / Banded ironstone	Rock gypsum / Rock anhydrite	Crystalline dolomite	
		coarse (>2.0 mm)	siliceous	ferruginous	salt minerals	calcareous	
				Composition of crystalline rocks			

TABLE 3.9
Common metamorphic rocks

Rock name	Common parent rock	Chief minerals	Identification features
Foliated			
Slate	Mudrock	Mica, quartz	Cleaves into thin plates; bedding planes of parent shale may show as lines on cleavage plates
Phyllite	Mudrock, tuff	Mica, quartz, garnet	Surfaces of plates highly lustrous; plates often sharply bent or wrinkled; garnet and other crystals on some plates
Schist	Mudrock, lava, tuff	Mica, quartz, plagioclase, epidote	Well-foliated with visible flaky minerals like mica, sometimes chlorite or talc; garnet and other crystals commonly present; foliations may be wrinkled
Gneiss	Granite, mudrock, rhyolite, etc.	Feldspar, quartz, mica, amphibole	Coarse-grained, with definite but imperfect foliation; bands made up of different mineralogical composition
Migmatite	Mixtures of igneous and metamorphic rocks	Feldspar, quartz, amphibole, biotite	Coarsely banded or veined; bands of highly variable mineralogical composition

Non-foliated			
Hornfels	Mudrock, slate, tuff, lava	Highly variable	Hard, massive and very fine-grained; breaks into sharp, angular fragments
Quartzite	Sandstone	Quartz	Very hard; interlocking quartz grains fused together; breaks pass through individual grains; colours white, grey, pink, red, purple but mainly light-coloured
Marble	Limestone, dolomite	Calcite, dolomite, silicates of Ca and Mg	Fine- to coarse-grained; often streaky and may contain calcite veinlets; effervesces with acid
Amphibolite	Gabbro, basalt, tuff	Hornblende, plagioclase, minor quartz and garnet	Often coarse-grained with linearly orientated laths of hornblende

The type of catenary relationship that dominates any particular landscape depends on the degree to which it has been dissected, on the extent to which the valleys have widened, and on the climate, parent material and age of the local planation surface. Under semi-arid or arid conditions where there has been no opportunity for deep leaching, the hillcrests may consist mainly of rock outcrop, and rock is entirely exposed in the free face (Fig. 3.6). Downslope from the free face, mass wasting processes often give rise to a talus slope which grades further downslope onto a gently sloping pediment surface mantled by a pediment deposit of fine colluvium.

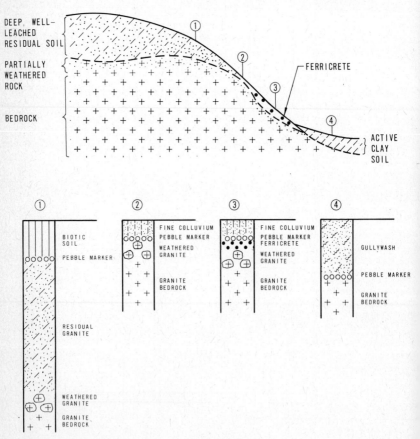

FIG. 3.5. A catenary sequence of soils developed on granite in a sub-humid environment.

Fluvial landscapes display similarly recurrent patterns of features such as river terraces, levees, backswamps, floodplains, and gully environments (Fig. 3.7).

In many of the extensive stable continental shield areas of the world which have not been subjected to Pleistocene glaciation, subaerial planation over the last 200 million years has produced a series of erosion surfaces. Depending on its stage of evolution, each local landscape is characterized by a typical suite of local landforms with associated soils and engineering properties. The pattern of such landscapes can usually be interpreted in terms of an idealized

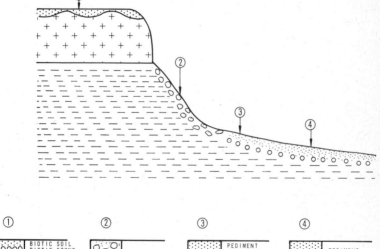

FIG. 3.6. The four elements of the hillslope in semi-arid and arid environments.

cycle of landscape development that begins with the uplift of an area (Fig. 3.2), its subsequent incision by the streams now raised above base level (sea level), followed by valley-widening, hillslope recession, and the formation of pediments. The latter eventually dominate the landscape and coalesce to form a multi-concave pediplain upon which residual hills and upland remnants are occasionally preserved. Many geomorphologists find it useful to divide this sequence into the stages of youth (deep incision of valleys into the original upland), maturity (maximum relief with valley widening) and old age (widespread plains often containing a few upland remnants).

The oldest of these cyclic landscapes pre-dates the major period of continental drift during the early Jurassic, 190 million years ago, and is usually restricted to a few isolated remnants which are of little local significance.

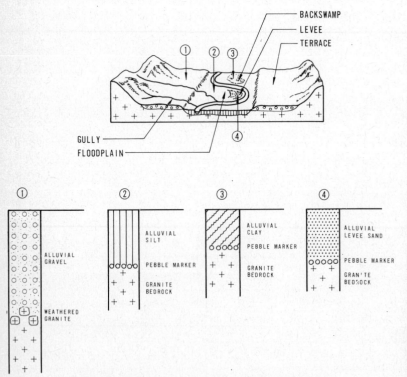

FIG. 3.7. Landforms associated with fluvial features.

A more widespread erosion cycle was initiated by continental disruption on a global scale at the end of the Jurassic—this cycle lasted for more than 100 Ma. The resulting erosion surface is in many places well planed and the underlying soil materials exhibit very characteristic engineering properties. All rock types were affected, although the deepest weathering mantles are encountered on igneous rocks owing to the generally greater chemical instability of the constituent minerals. In granite–gneiss terrains, such deep residual soils frequently possess a collapsible fabric, while basic igneous rocks such as plateau basalts produce deep clayey soils of high compressibility. Extensive pedocrete 'hardpans', formed under the prolonged influence of the shallow fluctuating water tables which characterize very flat areas, are commonly associated with remnants of this planation surface.

Another widespread erosion cycle of shorter duration was initiated during the mid-Tertiary. Those limited areas in which this cycle has achieved widespread planation show similar engineering characteristics to those of the mid-Cretaceous cycle. Elsewhere its encroachment on the older surface is associated with dissected topography in which landforms and soils display the sorts of relationships described earlier (Figs. 3.5, 3.6, and 3.7).

The landscape produced by these earlier cycles was subsequently incised, first by a late Tertiary erosion cycle and thereafter by rapid downcutting of the river valleys under the influence of localized tectonic movements which occurred · at the beginning of the Quaternary era. In such areas, which are often characterized by very rugged topography, pedogenesis has seldom reached an advanced stage except in humid environments, and the principal engineering problems are related to slope instability in deposits produced by mass wasting processes.

Many landscapes are multi-cyclic with the oldest, well-planed, surface preserved on plateau remnants and subsequent surfaces forming benches and pediments below intervening escarpments. The youngest cycle, which may be currently operating within the landscape, is often represented by deeply incised coastal gorges (Fig. 3.8). The recognition of the zones of each stage that separates an older from a younger erosion surface is vital in the mapping of catenas with different properties of materials and engineering problems.

Corresponding to the various erosion cycles described above,

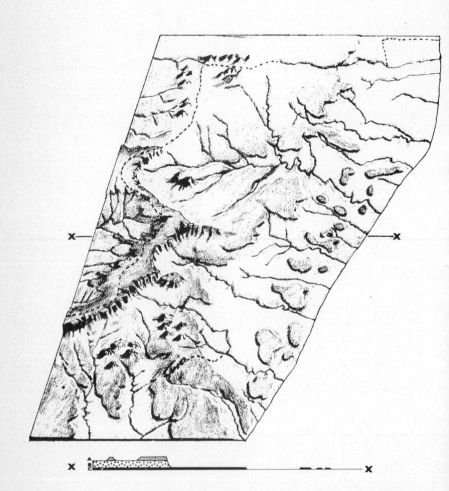

Fig. 3.8. A multi-cyclic landscape (after King 1972): several different erosion surfaces are represented, some in the form of small flat-topped residuals.

various depositional equivalents are preserved, chiefly in marine environments but sometimes also in continental areas. These take a variety of forms relative to the local geomorphological processes under which they were deposited, and the degrees of pedogenesis and lithifaction to which they have subsequently been subjected. Generally speaking, the products of older cycles exhibit a greater degree of lithifaction and pedocrete formation. However, these deposits present such a wide range of different materials that it is not possible to make meaningful generalizations about their engineering characteristics.

RECOMMENDED READING

Allen, J. R. L. (1970). *Physical processes of sedimentation*. Allen and Unwin, London.

Gass, I. G., Smith, P. J., and Wilson, R. C. L. (1972). *Understanding the earth: a reader in the earth sciences*, 2nd edn. Artemis, Horsham.

Holmes, A. (1965). *Principles of physical geology*, 3rd edn. Nelson, London.

King, L. C. (1962). *Morphology of the earth*. Oliver and Boyd, Edinburgh.

Leopold, L. B., Wolman, M. G., and Miller, J. P. (1964). *Fluvial processes in geomorphology*. W. H. Freeman, San Francisco.

McLean, A. C. and Gribble, C. D. (1979). *Geology for civil engineers*. Allen and Unwin, London.

Moseley, Frank (1979). *Advanced geological map interpretation*. Edward Arnold, London.

Platt, J. I. and Challinor, J. (1974). *Simple geological structures*. Murby.

Read, H. H. and Watson, J. (1970). *Introduction to geology*, Vol. 1: *Principles (and processes)*; Vol. 2: *Earth history*. Macmillan, London.

Sparks, B. W. (1979). *Geomorphology*. Longman, London.

Thomas, M. F. (1974). *Tropical geomorphology*. Macmillan, London.

Wilson, E. M. (1974). *Engineering hydrology*. Macmillan, London.

4. Soil data requirements for various engineering projects

Since practically all civil engineering works are built either on soil, in soil, or of soil, most require some form of soil investigation. The scope of the investigation depends not only on the complexity of the natural conditions but also on the requirements of the structure. The pattern of investigation for a housing scheme, which must consider the demands of lightly loaded structures, underground services, and streets, differs from that for a large power station with cooling towers, boiler-houses, and vibrating steam-turbines. 'If you do not know what you should be looking for in a site investigation, you are not likely to find much of value. What you look for should be suggested by the natural environment, and by the nature of the problem to be solved' (Glossop 1968).

The soil properties which bear upon the long-term performance of any structure are fundamental but consideration must also be given to temporary problems which may be encountered during construction, and also to problems that may arise during the lifetime of the project and long after the construction force has left the site. Soil conditions sometimes influence the whole design of the structure, e.g. on heaving clays it may be better to build a 'flexible' structure, by providing a number of properly constructed joints ('split-construction'), than to use continuous brick-walls, which are brittle and cannot withstand differential movement.

The objectives of a soil investigation may embrace any combination of the following (British Standards Institution 1957):

(i) to assess the general suitability of the site for the proposed engineering works;
(ii) to enable an adequate and economic design to be prepared;
(iii) to foresee and provide against difficulties that may arise during construction owing to ground and other local conditions;
(iv) to determine the causes of defects or failures in existing works and the remedial measures required;
(v) to advise on the availability and suitability of local materials for construction purposes.

There are several factors of general importance on which data should be sought at the beginning of the soil investigation for *any* type of project:

  (i) Any adverse effects due to the *water table* must be provided for in the design of a structure. Problems of heave, collapsible fabric or sinkhole development (Chapter 5) may be associated with a deep water table.

 (ii) Where the water table is shallow, there may be problems from surface seepage, or from compressible, sensitive, and active soils; this also presents problems in deep excavations.

(iii) Combinations of these problems may occur where there are major fluctuations in the level of the water table or perched water tables.

(iv) Sometimes the water table intersects the ground surface and gives rise to 'spring-lines' or seepage zones, which can cause trouble during excavation for shallow foundations or cause additional expense through the special measures required to damp-proof ground floors or basements.

 (v) The general layout of any scheme must also take the *drainage* of the site and the disposal of stormwater into account. In a major housing scheme in fairly flat country, for example, it is essential to ensure that the floor levels are at least 0.5 m above street level. In fact, the design of the stormwater drainage system should precede decisions on final levels for paved areas, parking grounds, or streets.

(vi) *Unstable slopes* can be a threat to any engineering works even if they are situated some distance from the site. Their state of delicate equilibrium may be upset, sometimes with catastrophic consequences, by the hydrostatic pressures which develop when water enters fissures or tension cracks in the soil mass, or by earthquake shocks.

(vii) *Seismic disturbances* can also cause certain types of saturated soils to liquify and flow, even on shallow slopes.

If it seems, in the early stages of the planning of any engineering project, that any of these hazards are possible, this will usually justify extending the investigation to predict their effects on the soil and the structure, e.g. unstable slopes and sensitive soils must be identified in areas of high seismic risk so that they may be avoided or rendered safe by appropriate treatment.

Various types of engineering project are grouped together below on similarities in their risk to human life or in the economic implications of their failure, in order to make generalizations about their soil data requirements.

URBAN DEVELOPMENT INVOLVING LIGHT STRUCTURES

Planning for urban development involves various types of land use and so must take into account a variety of natural constraints on siting, e.g. in a cemetery reasonably thick soils must be present above the water table. The major requirement of a soil survey for urban planning is to isolate specific problem areas and sources of useful construction materials such as brick clay, concrete sand, and road gravel.

Houses and other light buildings

Requirements

Much urban development comprises structures such as houses, schools, water-supply schemes, and sewage disposal works, which exert low pressures on the soil (less than about 100 kPa). Structures of this sort cannot usually tolerate differential movements of more than about 3 mm without the development of unsightly cracks or, in the case of water-retaining structures, intolerable leaks (Table 4.1). Furthermore they do not warrant the use of expensive foundations or structural treatment. Hence town planners should, where possible, zone the areas containing the best founding soils for house construction and place heavier structures, for which more expensive foundations are justified, on worse soils.

Soil conditions affecting these requirements

Both compressible and active soils are likely to give rise to unacceptable differential movements, the former as a result of consolidation under load and the latter as a result of shrinkage on drying out or expansion on wetting up. The soil beneath normal foundations (strip-footings, Fig. 4.1) must be capable of supporting

TABLE 4.1
Limiting deflection ratios in structures. (Modified from Burland and Wroth 1974)

Type of structure	L/H	Δ/L
Infilled and concrete framed structures	<3	0.002
	>3	0.002+
Load-bearing brickwork (ordinary dwellings,	<3	0.0003
plane brick walls, continuous brick cladding, etc.)	>5	0.0005

L is the length of the structure, H the height of the structure, Δ the maximum tolerable movement, and Δ/L the deflection ratio. For example, if a house is 12 m long and 3 m high the limit of movement before damage is about 5 mm.

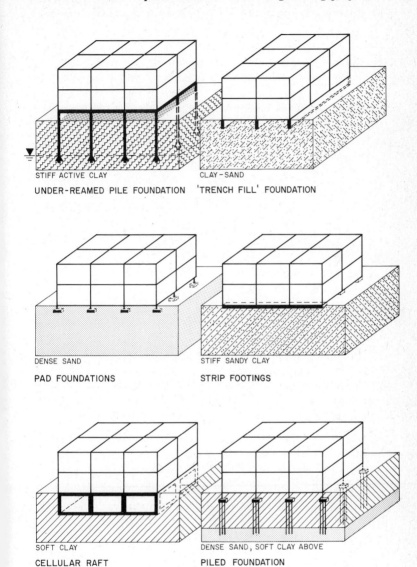

Fig. 4.1. Foundation types for light structures to suit differing soil conditions.

bearing pressures of about 35 kPa in the case of single-storey houses and 100 kPa in the case of double-storey houses, without undergoing excessive consolidation. On wetting up from a desiccated state, active clays may exert swelling pressures of more than 350 kPa which are able to lift parts of a lightly loaded structure; the vertical movements may reach 3 per cent of the thickness of the active clay layer. Conversely, when active soils dry out, shrinkage settlements may cause severe damage (Chapter 5). In temperate climates these may be firm shrinkable clays, while in arid areas stiff heaving clays are more common.

Solutions

The following foundation designs are suggested for houses and other small buildings or lightly loaded structures to be built on soil profiles for which various ranges of movement have been pre-calculated by the procedures outlined in Chapters 2 and 5:

(a) *Total vertical movement 0 to 5 mm (differential movement less than 2.5 mm)*
No special foundation treatment is required and structures may be founded on strip-footings of unreinforced concrete about 500 mm wide and about 500 mm deep (Fig. 4.1).

(b) *Total vertical movement 5 to 30 mm (differential from 2.5 to 15 mm)*
 (i) When produced by active soil at any depth within the soil profile: divide the building into several smaller units between which joints are left to permit relative movement ('split-construction'); this will avoid unsightly cracking as neat cracks have been built into the structure from the start, but are hidden from view behind cover strips or cornices, or are masked by other architectural features.
 (ii) When produced by shrinkage movements confined to the top 2 to 3 m of an initially wet and active soil profile: excavate narrow trenches and replace the unstable soil with concrete ('trench-fill', Fig. 4.1) in place of strip-footings.
 (iii) When produced by compressible soil: an alternative solution is to spread the load and thus reduce the bearing pressure on the soil by using wider strip footings.

(c) *Total vertical movement more than 30 mm (differential more than 15 mm)*
 (i) When produced by highly compressible soils: the mass of the building can be spread over its entire plan area by constructing it on a rigid raft, such as a reinforced concrete slab or a cellular raft (Fig. 4.1). In the most extreme cases use piles to transmit the applied loads to more competent materials in an underlying soil or rock horizon. Under-reaming, or belling out the base of the pile, reduces the bearing pressure if necessary.

(ii) When produced by swelling active soils above a deep water table: piles are founded either below the water table or in stable soil below the active horizon, and under-reamed to provide anchorage against uplift forces. The walls are constructed on reinforced concrete beams (grade-beams) which span between the pile caps, leaving a clear gap of at least 250 mm between the underside of the beam and the surface of the soil. Ground floors should be suspended above the surface of the soil on concrete slabs; where the cost of this cannot be justified, the floors must be kept entirely isolated from the walls by means of joints filled with a flexible sealant.

Much the same types of solution may be used for houses founded in soils with a collapsible fabric (Chapter 5). The additional settlement which accompanies the inundation of this type of soil when under load is usually localized (e.g. under foundations near leaking pipes) and therefore produces severe differential movement. Some alternative suggestions for foundation design are offered in Chapter 5. Yet other solutions have successfully been applied to houses built on these and other troublesome soils (Jennings and Evans 1962): one of the more novel of these is 'three-point suspension' which uses the principle that three points define a plane—a three-legged table does not wobble—and consists of three concrete pad-footings connected by rigid beams or a reinforced concrete slab, raised clear of the ground surface, upon which the house is erected. Hydraulic jacks may be used for the correction of any tilt resulting from differential movement between the three supports. Table 4.2 summarizes the soil data required for an urban development plan.

On any soil, localized accumulations of water under foundations should be avoided by draining rainwater away from the house; where a shallow or a perched water table causes surface seepage, agricultural or French drains (perforated pipes laid in a bed of gravel protected by filter material) may be used to remove water to the lowest point of the site.

Minor roads and streets

Requirements

A road or street is made up of several layers of soil, stone, bitumen, or concrete; its function is to carry traffic smoothly and safely in all weather. The configuration of these layers, which together make up the *road prism*, is shown in Fig. 4.2. The uppermost is the *surfacing* which is either flexible (when bituminous) or rigid (when of concrete) and must be durable, skid-resistant and waterproof to protect the lower layers of the prism from the entry of water. The next layer is the *base* which consists of a natural gravel or a crushed rock and provides a strong but flexible support for the surfacing. It must be strong enough to support the required wheel-loads and thick enough to spread them more uniformly over the weaker layers beneath. The *subbase* consists of selected granular soil and further

TABLE 4.2
Soil data required for an urban development plan

  (i) Good founding material (non-active soils of adequate bearing capacity): strip footings;
 (ii) Active clays, with swelling or shrinking potential: houses on soil of medium potential expansiveness may require 'split-construction' while piled foundations may be required in soils of high potential expansiveness;
(iii) Soils with 'collapsible' fabric: if shallow (less than about 2 m) they can be compacted or completely removed, if thick they require deep foundations;
 (iv) Compressible soils: shallow layers may show only minor settlement potential, but areas of large settlement potential require special measures;
  (v) Reclaimed areas, particularly old refuse dumps;
 (vi) Areas prone to large-scale subsidence, e.g. potential sinkhole zones in karst terrain, or areas undermined at shallow depth (less than about 200 m);
(vii) Solid rock outcropping at surface or present at a depth of less than a few metres;
(viii) Unstable slopes or areas where landslides, or solifluction, have occurred;
 (ix) Areas of high groundwater table or seepage zones, particularly where corrosive soils may lead to sulphate attack on pipes;
  (x) Flooding, e.g. the fifty-year flood-line of local streams;
 (xi) Useful natural resources, e.g. clays for brick-making; sands for plaster, concrete, filters; gravel for roads; rock for quarrying.

---

spreads the traffic loads over the weak fill or natural soils *in situ* which comprise the *subgrade. Fill* is used to cover *in situ* soil where this is of low strength or high compressibility, or where the road prism must be raised above the level of a shallow water table, or where a particular *grade* or elevation must be attained, e.g. across valleys, over culverts or against bridge abutments. The overlying layers dissipate the traffic loads, so that the requirements for fill are much less rigorous, but it is none the less important not to use active clays or soils which have a low shear strength even after moderate compaction. Fills are not much used in the construction of minor roads and streets in most urban areas.

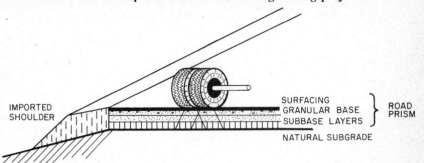

FIG. 4.2. The road prism is a structure to spread the load so that it can be carried on the natural ground or subgrade.

Soil conditions affecting these requirements

The properties of the natural subgrade soils have the most influence on the design of the road prism, since they dictate its total thickness. Active clays, highly compressible or sensitive soils and materials which tend to soften markedly with the build up of moisture below the road prism make poor subgrades.

Solutions

The adverse effects of unfavourable subgrades are counteracted by increasing the thickness of the prism: in some cases this may have to exceed 500 mm, whereas with a well-drained granular soil it may not be necessary to do more than compact the natural material *in situ* and cover it with say 150 mm of base before applying the surfacing. Where the subgrade is permanently or seasonally water-logged or drains slowly after rain (as for example in cuts in rock), subsurface drains are installed to protect the prism from repeated saturation, which weakens the base, may lead to its distortion, and ultimately to failure of the surfacing.

The thickness of the individual layers depends on the properties of construction materials from local sources and may dictate the need to use stabilizing agents such as lime or cement. In some cases it may be necessary to blend materials from different sources to achieve the appropriate grading. Potential construction materials that contain soluble salts in deleterious concentrations (Chapter 5) may have to be specially treated or avoided.

The Unified and PRA Soil Classifications greatly assist the assessment of soil for *in situ* subgrades or construction materials (Tables 2.2 and 2.3).

There are several design methods for determining the required strength of each layer of the road prism. They set lower limits to the quality of the material to be used in each layer, usually in terms of grading and plasticity index. In the CBR method the thickness of the various layers is related to the CBR strength (p. 276) of the construction materials by means of a set of graphs (Road Research Laboratory 1961). Other design methods relate empirically the contribution of each particular type of material to the overall strength of the prism.

The final design of the road is based (pp. 93–4) on a preselected design life and intensity of use. A typical requirement for a main road on which the traffic is increasing at a rate of 6 per cent per annum is:

Design life: 20 years
Design wheel-load: 40 kN
Number of repetitions of wheel-load during design life: 300 000

Fig. 4.3 illustrates alternative road designs which would meet these traffic requirements. In the first example, the subgrade requires a total thickness of 670 mm using unstabilized material; in the second example, a thickness of only 500 mm of cement-stabilized material is required. The relative merits of these alternatives would have to be evaluated in the light of their respective costs.

Typical specifications for the selection and treatment of natural construction materials for roads and embankments are given in Table 4.3.

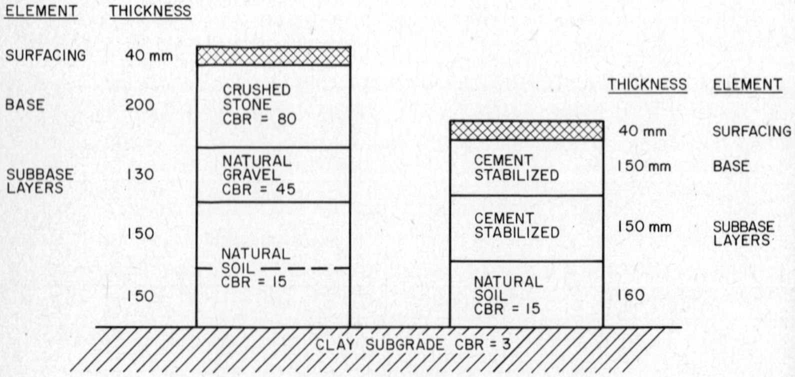

FIG. 4.3. Alternative designs for a road prism.

TABLE 4.3

Typical specifications for the elements of a road prism

	Fill	Lower subbase	Upper subbase	Base (natural gravel)
Min. mod. AASHO density	90	93	95	97
Min. UCS (kPA)			750	1000
Min. CBR	3	15	45	80
Max. PI (unstabilized)		15	10	10
Max. PI (stabilized)		10	6	6
Max. size of particles (mm)	600	100	100	100

Underground services in trenches or shallow tunnels

Requirements

Any residential community requires an extensive network of essential services such as pipes or conduits carrying fresh water, stormwater, sewage, gas, and electric cables. Most of these are placed underground for aesthetic and economic reasons: since trenching and backfilling is cheaper than tunnelling the latter is only used as a last resort, usually for the installation of services beneath existing buildings. Accommodating all these utilities requires a knowledge of soil conditions to a depth of at least 1.5 m but sometimes 3 m or more (Fig. 4.4).

Soil that is easy to excavate, and that is stable in the sides of trenches during excavation and installation, facilitates the placement of services. Pipes should not leak since this may cause pollution and wetting of the soil, and, leakage *into* the pipes may cause overloading of a stormwater system or additional costs in the processing of an unnecessary sewage load; there should thus be no appreciable movement of the pipes, which might disrupt connections, nor should pipes be distorted through localized overloading. Corrosion may also be a major hazard.

Soil conditions affecting these requirements

Soil conditions which affect these requirements are: a high water table which hinders excavation and reduces the stability of trenches or shallow tunnels; soft or firm soil which facilitates excavation and, conversely, rock or pedocretes which are difficult to penetrate; soils of low shear strength which are unable to stand unsupported; active and compressible soils which are subject to differential

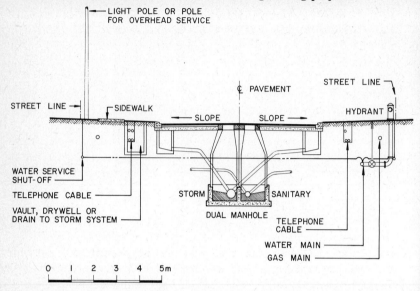

FIG. 4.4. A variety of services is buried beneath most urban streets. (After Anderson 1973.)

movements on changes in moisture content (Chapter 5) and may disrupt connections; karst conditions in which leakage of water may cause instability (Chapter 5); frozen soil conditions which may preclude burial of services (Chapter 5); the nature and degree of compaction of trench backfill which can damage pipes if unsuitable or can function as a French drain; and the presence of soluble salts and aggressive groundwater in the surrounding soils or in the trench backfill which may corrode pipes or cables.

Solutions

Pumping and support of trench sides is imperative where the water table is shallow. Since the major cost for small diameter pipes is in excavation and back-filling, the choice of an alignment which crosses the most easily excavated soils may produce major savings. Planes of weakness in the soil such as fissures and slickensides, and soils of low cohesion such as sands, invariably require the shoring of trenches; shoring is normally required in any case when the depth of the trench exceeds 1.5 m. Shallow tunnels in soil always need temporary support during construction and often require a permanent lining. Water is always a danger in excavation and may cause 'running sand' conditions, or softening of clays and silts, so that the lateral pressures on shoring and bracing increase with time, sometimes beyond the capacity of supports that were initially adequate.

If pipes are laid in soft compressible soils which are affected by nearby foundations, or on active clays, they may be subject to differential movements. For rigid earthenware, concrete or iron pipes, these movements are best accommodated by flexible joints between shorter pipe lengths. The flexibility of steel pipes with welded joints may be adequate. In very soft soils compaction of the bottom of the trench may limit differential settlements.

Backfill of boulders and coarse gravels should be avoided since they tend to exert point loads which may damage the pipe. Ease of compaction is important or a backfill which is looser than the surrounding soil may subside, but over-compaction may damage the pipe. Ideally, pipes are laid on a bed of selected material, such as single-sized (e.g. 10 mm) free-running crushed stone or a smaller 'pea gravel', and then covered with selected and compacted backfill; if the French drain so formed is likely to cause problems, as in karst areas, care must be taken to provide cut-off seals at regular intervals.

The presence of soluble salts or aggressive groundwater must be identified (Chapter 5). Where they are present, concrete pipes of sulphate-resistant cement, or bituminous or epoxy coatings on metal pipes, may help to prevent corrosion. Where the problem is severe the use of selected salt-free backfill is desirable. Care must also be taken not to incorporate ferruginous concretions which promote electro-chemical corrosion.

MAJOR EARTHWORKS

Requirements

The construction of transportation routes in areas of even moderate relief requires the excavation of cuts through ridges and the placing of fills through valleys, so that the gradient of the route suits the type of traffic. For railways the maximum gradient is often limited to 1:66 while in major roads the maximum is about 1:15. The horizontal curvature cannot be too pronounced if high speeds must be accommodated, and the vertical curvature, or rate of change of grade, must always permit adequate sight distance. For airfields, large areas must be shaped to give a fairly level surface but yet be so designed as to afford adequate runoff and drainage. Thus geometric constraints often demand that large volumes of earth be moved, and the alignment of routes or the location of runways will be governed, at least partly, by the economics of these earthworks (Fig. 4.5).

Major earthworks also include excavations for cuts, fills and bridge embankments and the materials selected for the controlled layers of the road or runway prism or the railway formation, which support the traffic loads. Poor materials may have to be discarded as spoil, and better materials imported, but the engineer always

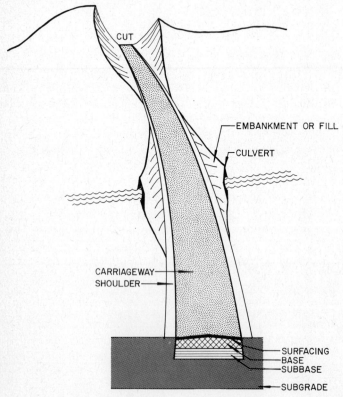

FIG. 4.5. In areas of high relief, transportation routes require major earthworks, including cuts and fills.

tries to balance cut and fill by using any suitable materials traversed by the routes. Canals consist either of cuts or of parallel twin fills or levees; they are particularly sensitive to leakage through permeable subsoils, or to differential movements that rupture the embankment or lining.

Cuts must be permanently stable under all weather conditions and they should be easy to excavate. Fills must not suffer slips owing to shear failure and must not settle to such an extent that they develop major undulations or suffer cracking.

Soil conditions affecting these requirements

In cuts the shear strength of the exposed faces dictates the maximum stable *batter* or slope angle. The stronger the materials the steeper will be the batter, the smaller the volume to be excavated and the shorter any overpass bridges. Low shear strength presents severe problems in old landslide areas and in clayey soils, particularly if they are highly fissured or slickensided. Any change in the configuration of a natural slope by the construction of a highway may upset delicate equilibria in areas disturbed by ancient landslides (Fig. 4.6); in such slide areas minor changes in soil moisture may trigger severe instability. Areas of 'slide topography' are indicated by slumps, scars, hummocky ground below scarps, leaning trees, or displaced fences. Slope failures often result from changes in the level of the local water table, so indications of a fluctuating water table should be sought even if the conditions are dry at the time of the survey. In Fig. 4.7 pore-water pressure caused failure along a slip surface when a tension crack caused by small movements near the top of the cut became filled with water.

Poor subgrade conditions, and especially compressible, active and sensitive soils, affect the performance of fills (Fig. 4.8).

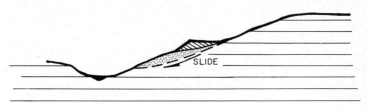

FIG. 4.6. Ancient landslides may be reactivated if an embankment is placed upon them unknowingly.

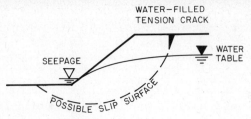

FIG. 4.7. Hydrostatic pressure, either below the water table or in a water-filled crack, can affect the stability of a cut or a natural slope.

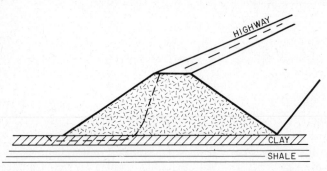

FIG. 4.8. An embankment may fail if there is an underlying layer of soft clay.

Solutions

The road, railway, or canal route is best located to avoid unstable slopes or areas in which rock is present at shallow depth. Where the route for a road or railway must cross unstable slopes the stress on the soil can be reduced by decreasing the height of a fill or the depth of a cut, or by reducing the slope of the excavation or providing additional subsoil drains. Canals are best routed through clayey soils of low permeability; alternatively, they may be lined with a clayey material (Fig. 4.9).

The erosion of cut slopes is controlled by providing cut-off drains above the cut and at the foot of the slope. In every case a cut must be designed to take into account local soil conditions which may affect its stability. In weak or fissured materials drains to relieve high water pressures may suffice, but batters as low as 2:1 (horizontal:vertical) may be necessary, often in conjunction with benches; in extreme cases the slope may have to be supported by

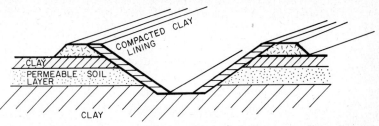

FIG. 4.9. Local clayey materials may be used to line a canal to prevent excessive seepage into a permeable subgrade; a clayey material with a PI in the range 10 to 12 and a LL not exceeding 45 is most suitable (Parry 1978).

cable-anchors or retaining walls. The shear strength along some fissures is far lower than that of an intact mass, and the investigator should note the orientation, spacing and continuity of clay seams of low strength within a harder mass. Fig. 4.10 illustrates failure along discontinuities in the soil, which could have been avoided by flattening the batter or by tying back the unstable wedge with anchors extending to behind the slip planes. Cut faces in soils which are susceptible to erosion and sloughing must be protected by grassing or masonry.

Deep fills should not be constructed on weak or compressible subgrade soils and where these cannot be avoided the site may be pre-loaded. This entails the placing of a surcharge load, such as a high fill, which applies a pressure greater than that to be exerted by the final structure, and which leaves the soil in an over-consolidated state when it is later removed (i.e. the soil is permanently deformed

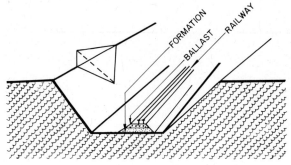

FIG. 4.10. Unfavourable orientation of fissures in a clay may cause wedge failure in a cut.

by applying a temporary preconsolidation stress greater than the final design stress). This technique may be effective where the approach fills to a bridge over a soft silt are built higher than will be required in the finished construction, but it may take too long to consolidate a clay adequately unless the application of the surcharge is supplemented by installation of sand drains. Where necessary, adequate underdrains filled with clean sand and gravel must be provided to prevent the build-up of water pressures which could threaten the stability of the embankment. While the specification of the quality of materials for fills is not nearly as rigorous as for the controlled layers which they support, active materials and dispersive soils should be avoided (Chapter 5).

'Reinforced earth' may be used in areas of poor ground, or where there are constraints on the space that may be occupied by the side slopes of embankments, by building the bank in layers behind an articulated wall, each section of which is anchored by a strip of metal extending back into the fill. Construction is fairly simple, involving the placing of precast wall sections, instead of casting concrete in place, and the strips are placed on top of successive layers of fill and do not interfere with normal compaction (Fig. 4.11). Careful design is required, taking account of the possibility of circular slip, for overall stability.

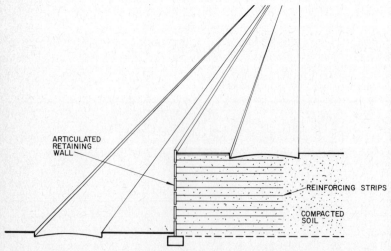

FIG. 4.11. Reinforced earth offers an alternative to ordinary retaining walls.

If the controlled layers of a road or runway are to be placed directly on a subgrade of natural soil, the route should be selected to avoid traversing areas occupied by active, marshy, weak, or compressible materials. In some cases, poor subgrades can be excavated and replaced by compacted material or mechanically stabilized by special methods, e.g. by using an impact roller or by dynamic consolidation (p. 97); in others, chemical additives may be more effective. Where possible the route must lie close to proven sources of construction materials (Chapter 9). The combined thickness of the various controlled layers will depend on the traffic loads to be carried and the properties of the subgrade as above (p. 84); Table 4.3 is a typical specification for materials for road construction, based on the CBR method of control. Numerous indicator tests are used to classify soils between CBR sampling points. An alternative design method is based on the modulus of subgrade reaction, values of which are determined in the field by plate-loading tests (Road Research Laboratory 1961). Table 4.4 summarizes the requirements for materials for a railway formation.

For roads and runways the most suitable material for the upper

TABLE 4.4

Guide to the selection of formation material for railway construction.
(After Rauch 1963)

Material	Unified soil class (Table 2.2)	Soil property	Usage
Type A	GW, GP, GM, GC, SW, SP, SM, SC	Liquid limit: less than 40% Plasticity index: less than 20% Linear shrinkage: less than 8% Less than 75 μm size†: less than 50%	Within 1 m of formation level‡ in cuts and within 2 m in banks
Type B	ML CL OL	Liquid limit: 40–55% Plasticity index: 20–25% Linear shrinkage: 8–10% Less than 75 μm size†: 50–70%	In lower parts of banks up to 2 m below formation level
Type C	MH CH OH	Liquid limit: more than 55% Plasticity index: more than 25% Linear shrinkage: more than 10% Less than 75 μm size†: more than 70%	Not permitted unless stabilized with lime or cement

† Indicates percentage of silt and clay in the soil.
‡ The top of formation is immediately below the ballast.

controlled layers is a well-graded gravel with a binder of low plasticity, which is easy to compact. In a railway the ballast, or upper layer, is exposed between the sleepers; it must, therefore, be more durable than a road or runway base and commonly consists of single-sized crushed stone. The prism of a major road incorporates about 10 000 m^3 of selected construction materials of high quality per kilometre length; this volume does not include any additional fill material to maintain the grade in undulating terrain.

If local materials in their natural state are unsuitable for the controlled layers, their properties can be improved by additives such as lime or cement. Lime is generally used to reduce the plasticity of clayey soils and to inhibit volume changes on fluctuations in moisture content; the addition of, say, 4 per cent of hydrated lime may reduce the plasticity index of a clay from about 20 per cent to 8 per cent, and it also combines with the clay to give the soil a higher strength after compaction.

Cement is used chiefly with sandy soil, and cements the grains to give the soil a strength equivalent to that of an interlocking gravel; the addition of about 5 per cent cement can often increase the unconfined compressive strength of a sandy soil to about 1500 kPa. With both these additives there are problems of shrinkage on drying out, even when the curing process, or setting of the cement during continual wetting, is carefully controlled. In arid areas much water may also be lost through evaporation during the mixing process, and soluble salts may become excessively concentrated; chlorides and sulphates affect the cementing action of both cement and lime and also damage the surfacing.

The provision of water for compaction and other purposes may present major problems in arid areas and may necessitate special exploration programmes to locate underground sources.

MAJOR STRUCTURES

Requirements

Major structures include high-rise buildings, power stations, factory plants, silos, bridges (including road or rail interchanges), tall chimneys and masts, elevated water towers, harbour works, and concrete dams. Large earth and rockfill dams are considered separately below.

The three main requirements of all major structures are that:

(i) they must not tilt or collapse due to shear failure in the underlying soil or fractured rock;
(ii) they must not settle unduly (Table 4.1);
(iii) they must not cause distress to nearby structures because of strains which they produce in the contiguous soil mass (Fig. 4.12).

In the case of harbour works and bridges, the soils beneath the foundations or retained behind any parts of their structures must not be affected by wave or current scour.

The foundations of major structures must be designed in relation to the constraints imposed both by the nature of the structure itself and by the soil; some structures can tolerate significantly more differential settlement than others without experiencing distress (Table 4.1).

Soil conditions affecting these requirements

The local soil conditions which affect these requirements extend to the depth of stress influence of the structure, i.e. to at least 1.5 times its minimum plan dimension, or to competent rock if this is present at shallower depth.

The shear strength of the soil, which gives its ultimate bearing capacity, is vitally important to very heavy structures or structures subject to high transient wind loads. If the foundation pressures exceed the shear strength of the soil, catastrophic tilting and collapse may take place, but this type of failure is relatively uncommon since foundation pressures of this magnitude would usually produce unacceptably large settlements due to consolidation of the soil, which would have been avoided in the original design.

Pressures which are insufficient to cause significant plastic strains may still lead to settlement on compressible soils, which causes distress such as cracking or misalignment. The likely extent of consolidation and the rate at which it will take place are both important; consolidation may continue over many years in clayey soils of low permeability, whereas on sandy and some silty soils it ceases soon after the load has been applied.

The ease of excavation of deep basements or other subsurface installations is also important. If the blasting of bedrock is required to excavate a basement in a built-up area the likelihood of damage to neighbouring structures might lead to modification of the design.

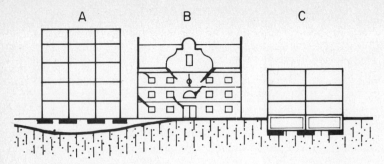

A   B   C

COMPRESSIBLE SOIL

Fig. 4.12. Consolidation of soil under a new structure may cause distress in the adjacent buildings.

The oldest building (B) was built of load-bearing brickwork and most of its settlement took place during construction; the final finishes were applied only after equilibrium had been reached. The structure on the right (C) was founded on a deep basement so that the net loading on the soil did not change significantly and little or no consolidation settlement resulted. The newest structure (A) of infilled framed reinforced concrete is relatively rigid, so that it acted as a monolithic block and suffered no differential settlement itself. However, the building in the centre (B) was 'dragged down' in the settlement bowl caused by the additional loading on the adjacent site and, being unable to resist distortion, was badly cracked.

The erection of a new structure adjacent to building (C) should not cause similar difficulties because the deep basement forms a very rigid box foundation which can withstand distortion, but underground services might be disrupted by the application of additional loads.

The photograph shows the pattern of cracking suffered by building B.

A shallow water table likewise imposes constraints on the excavation of deep basements.

Sands and silts of low cohesion are susceptible to erosion by wave and current scour in the vicinity of harbour works and bridges, and scour effects may persist to great depth.

Solutions

The engineering solutions to adverse soil conditions depend on the nature of the structure to be erected. If the calculated settlement exceeds what the structure can tolerate, the load may have to be applied to less compressible material at greater depth, by piling (Fig. 4.13). If piles are taken down to hard strata it is important to ensure that no softer layers underlie them. Another possibility is to provide a 'floating foundation' like a cellular raft, through which the total weight of the structure counterbalances the weight of the soil removed (Fig. 4.1).

Alternatively, the compressible soil can be treated by *in situ* densification to reduce the potential settlement, for example by:

 (i) dynamic consolidation, in which a heavy mass is dropped repeatedly on to the soil from a considerable height (say 35 tonnes dropping 30 m): this may achieve compaction to depths of up to 15 m below surface;

 (ii) vibratory treatment, in which the soil is agitated to the required depth by jetted water accompanied by mechanical vibration of the mass and addition of material to induce a denser state of packing;

(iii) grouting, in which a self-hardening compound or two interactive chemicals are injected successively into the voids.

Sometimes it is more economic to replace the compressible soil with a more suitable imported material which is compacted in layers, if necessary with the addition of stabilizing agents, to form an 'engineered fill'.

Where scour presents a problem, structures are best supported on piers, driven piles or caissons (i.e. large diameter steel or concrete tubes) founded on rock; alternatively sheet-pile walls can be used to retain and protect erodible soils, or *dolosse* (i.e. interlocking concrete armouring units) are placed to protect the surface. The stabilization of erodible materials by grouting is less satisfactory.

Where foundations have been designed to accommodate transient loads, as in water-retaining structures or in high masts which are subject to periodic wind-loading, uplift forces will occur from

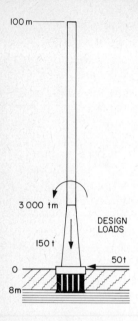

100 m

3 000 tm

DESIGN
LOADS

150 t

0

50 t

8m

BORED UNDER-REAMED PILE
ANCHORAGE OF CHIMNEY STACK

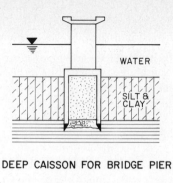

WATER

SILT &
CLAY

DEEP CAISSON FOR BRIDGE PIER

ANCHORAGE FOR PYLONS ON TRANS-
MISSION LINES, THROUGH DRIVEN
CAST-IN-PLACE PILES

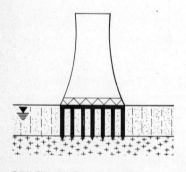

DRIVEN PILES TO ROCK

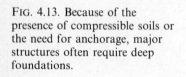

FIG. 4.13. Because of the
presence of compressible soils or
the need for anchorage, major
structures often require deep
foundations.

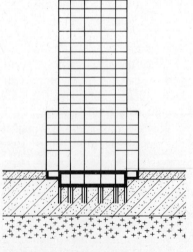

BASEMENT CONSTRUCTION RELIEVES
PRESSURE ON SOIL
(AFTER TOMLINSON, 1963)

time to time over part of the foundation; these can be counteracted by anchoring against uplift by cables grouted into competent rock, or piles under-reamed to provide widened bases in the soil.

Deep basement excavations, especially in urban areas where the use of shallow-angle batters is precluded by the presence of adjacent structures, frequently require support by cable-anchors, sheet piles, or perimeter piles between which shoring is installed (Fig. 4.14).

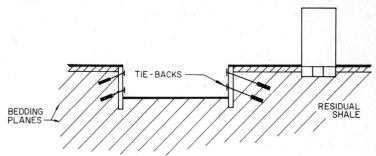

FIG. 4.14. Anchorage of the sides of a deep excavation depends on the disposition of planes of weakness in the soil.

Excessive inflow of water into such deep excavations may be countered by several treatments, e.g.

(i) the provision of drains and sumps or well-points to lower the level of the water table locally during excavation: drains usually consist of sand or gravel filters in trenches of appropriate depth, and they discharge into sumps from which the water may be pumped; well-points consist of perforated pipes, covered by a filtering screen of mesh, which are jetted or drilled into permeable soil materials and through which the water table may be lowered by pumping;

(ii) the injection of cement or chemical grout to form an impermeable curtain below a dam wall or around an excavation;

(iii) the construction of diaphragm walls† or of sheet piles which provide a continuous impervious barrier to the required depth;

(iv) the local freezing of the soil immediately behind an excavation face for a period sufficient to allow the installation of the structure or of more permanent drainage measures: this is accomplished by circulating a refrigerant, commonly brine, through pipes inserted in the side of the excavation.

---

† A diaphragm wall is a continuous wall constructed below ground level and is usually made by excavating a trench and filling this with concrete. To prevent the collapse of the sides during and immediately after excavation, the trench is often filled with a bentonite suspension which is later displaced from the 'slurry trench' by concrete to create the final diaphragm wall.

EARTH DAMS

Requirements

Water-retaining embankments are used in earth dams and also to form protective dykes around the perimeters of polders. Their water-retaining capacity is imparted by a relatively impermeable zone or 'core'. Usually this is of clay and the outer zone or 'shell' is composed of permeable materials if these are locally available (Fig. 4.15). Permeable materials are granular in texture and have a higher shear strength than the core material which they support. The shell is, in turn, protected against wave erosion by a covering of rock fragments known as 'rip-rap' or by concrete blocks. A horizontal filter-blanket is often incorporated to intercept any potentially damaging seepage which reaches the down-stream face of the dam; other filter zones may be used elsewhere in the dam. If

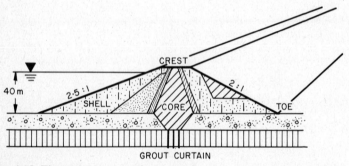

FIG. 4.15. The various zones of a composite earth dam require materials of different properties.

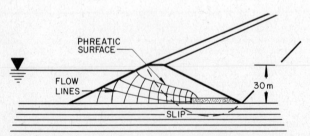

FIG. 4.16. Seepage through a homogeneous earth dam must be controlled by proper drainage measures so as not to endanger the stability of its downstream slope.

the local soil materials are preponderantly fine-textured the embankment may be homogeneous (Fig. 4.16).

In areas where rock is more abundant than granular soil the outer shell may be constructed of broken rock and constitutes the bulk of the embankment; a transition zone of intermediate grading separates the core from the rockfill and prevents the clay from intruding into the voids in the rockfill (Fig. 4.17). Where no clay is available, the impermeable core may be replaced by a concrete or bituminous membrane on the upstream face of the rockfill. Spillway channels for dams are generally of concrete.

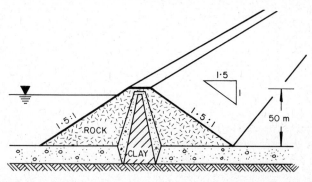

FIG. 4.17. A rock-fill dam may have relatively steep slopes. The clay core is surrounded by a suitable transitional material of intermediate grading to prevent the clay from infiltrating into the voids within the rock-fill zone.

Whatever the combination of materials used, the most important requirements are that the dam should not be washed away, or suffer shear failure or leak, even under full discharge conditions during maximum flood or during sudden drawdown on emptying.

Soil conditions affecting these requirements

The properties of the soil which most significantly influence the performance of an earth dam are the permeability of the sub-foundation materials and of the compacted core (Sherard, Woodward, Giziensky, and Clevenger 1963); the shear strength and compressibility of the sub-foundation materials; the presence of dispersive clays which can cause 'piping' (Chapter 5); and the zoning of the embankment materials which affects the configuration of the phreatic surface (Fig. 4.16).

Solutions

In order to control seepage through soils beneath the embankment a deep 'cut-off' is often provided in the form of an extension of the clay core into a trench which is excavated down to less permeable materials. Other solutions include the use of permanent sheet-piles or some other form of impermeable cut-off such as a diaphragm wall or concrete barrier. Despite the fact that the dam is usually constructed inside a temporary coffer-dam around which the river flow is diverted, the excavation of an open trench in the river bed for the installation of a cut-off may be difficult due to rapid seepage. If pumping from well-points is unable to stem the flow, temporary sheet-piles or diaphragm walls may be required to protect the sides of the excavation. Beneath the cut-off, less permeable materials such as fractured rock may have to be sealed by grouting along one or more lines parallel to the axis of the embankment.

Since earth embankments are particularly vulnerable to cracking due to consolidation of the sub-foundation soils, it is imperative that weak materials which may cause such distress are removed before construction so that the embankment is founded directly on materials of adequate strength.

TABLE 4.5

Guidelines to the acceptable ranges of properties for materials used in the zones of a composite earth dam

Soil parameter	Impermeable core	Semi-permeable intermediate zone	Permeable shell
Grading	Fine	Medium–coarse	Coarse
Clay (%)	10–30	5–10	<5
Liquid limit	25–40–60	<25	<20
Plasticity index	10–30	<10	<5
Linear shrinkage	6–14	<5	<2
Optimum moisture content (%)	12–25	10–15	8–12
Modified AASHO dry density (kN m^{-3})	14.0–16.5	15.5–17.5	16.5–19.0
Angle of shearing resistance ϕ' (degrees)	20–30	30–35	>35
Cohesion c' (kPa)	25–50	25	<25
Permeability (m s^{-1})	1×10^{-9}	1×10^{-7}	1×10^{-5}

In broad terms, the materials for each zone of the embankment must not be dispersive, must compact well and must have an adequate shear strength after compaction. Table 4.5 provides a general guide to the selection of materials for the various zones of the embankment. In practice the engineer is often forced to adapt his design to accommodate the materials available close to the dam site. The provision of adequate drainage in the embankment is equally important, and suitable sources of clean sand or fine gravel must be located for this purpose.

The thickness of the various zones and the mode of compaction of their constituent materials also require careful consideration. Hydraulic fracturing can occur in narrow clay cores owing to the arching of the clay between the less compressible shell or intermediate zones. An excessive hydraulic gradient in a narrow core may result in internal erosion or 'piping'. These phenomena are thought to have contributed to the failure of the 93 m high Teton Dam in Idaho, USA, on first filling. The loss of 14 lives and damage estimated at $400 million emphasizes the responsibilities of engineers engaged in such projects.

It is common practice to install instruments in earth dams and other large embankment structures to monitor their long-term performance. Piezometers measure changes in pore-water pressure, and earth pressure cells or other devices measure displacements in the embankment. Their readings provide invaluable information on the performance of the structure, and case histories based on them lead to improved design concepts. In particular, piezometer readings indicate the rate at which pore pressures dissipate during construction and hence the maximum rate at which the embankment can be safely raised. Piezometers also indicate fluctuations in pore water pressure during drawdown of water in the dam; these fluctuations affect the stability not only of the embankment but also of any natural slopes in the dam basin which are prone to landslides.

RECOMMENDED READING

Useful guidance on preliminary work for civil engineering may be obtained from a series of Civil Engineering Codes of Practice published by the British Standards Institution which represent 'a standard of good practice and take the form of recommendations'. This series includes the following:

CP 2001, Site investigations; No. 2, Earth retaining structures: CP2003, Earthworks; No. 4, Foundations.

Tomlinson, M. J. (1963). *Foundation design and construction*. Pitman, London. Includes many examples of the applications of soil mechanics to foundation engineering for buildings or structures.

Sherard, J. L., Woodward, R. J., Gizienski, S. F., and Clevenger, W. A. (1963). *Earth and earth-rock dams*. Wiley, New York.

5. Engineering soil problems

Chapter 4 discussed the problems and requirements of particular groups of construction; this chapter discusses particular groups of problems. They can often be predicted from a knowledge of the soil's origin. Therefore, before embarking on an exploration programme (Chapter 6), the observer should try to predict the origins of the soils likely to be present at a site, by means of geological or soil maps, geotechnical data banks etc. and surface observations, to guide his choice of field equipment and investigation procedures. Some of the most commonly occurring types of engineering problem are considered below and related to the depth of the water table.

PROBLEMS ENCOUNTERED IN AREAS OF SHALLOW WATER TABLE

Compressibility

Compressible soils are liable to consolidate under applied loads, leading to settlement. This type of settlement is commonly associated with transported soils, e.g. recent alluvial, lacustrine, and marine deposits, most of which have not been significantly desiccated or compressed by temporary loads such as those imposed during glaciation or by deep sedimentary mantles which have subsequently been removed by erosion, and so they have been consolidated only under their own overburden pressure, i.e. normally consolidated.

The leaning tower of Pisa is a classical example of the differential settlement of a heavy structure on saturated compressible soils (Fig. 5.1). This structure is 60 m high and has a total mass of some 14 450 tonnes, which exerts a mean bearing pressure of 480 kPa over a base width of 20 m. It is founded on about 30 m of marine clay which contains horizons of sand. The coefficient of consolidation c_v averages about 1.6 m^2 per year. Construction started in 1173 and was completed in 1317. Since then there has been a total settlement of some 2.5 m with a differential movement of 1.85 m

FIG. 5.1. The Tower of Pisa leans because it was founded at shallow depth on compressible clayey soils intercalated with permeable sandy horizons; the thickness of these varies laterally. Beginning of construction 1173; exploration by Terracina 1962.

DATUM LEVEL: BASE OF TOWER WHICH IS 2.4 m BELOW GROUND SURFACE

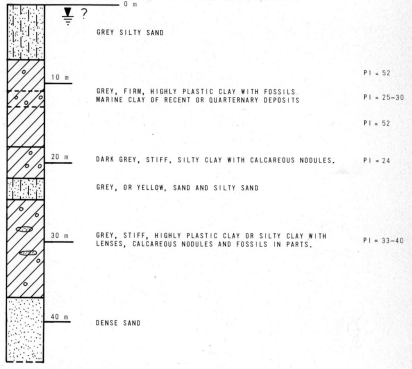

— 0 m

▼ ?

GREY SILTY SAND

— 10 m

GREY, FIRM, HIGHLY PLASTIC CLAY WITH FOSSILS.
MARINE CLAY OF RECENT OR QUARTERNARY DEPOSITS.

PI = 52

PI = 25-30

PI = 52

— 20 m

DARK GREY, STIFF, SILTY CLAY WITH CALCAREOUS NODULES.

PI = 24

GREY, OR YELLOW, SAND AND SILTY SAND

— 30 m

GREY, STIFF, HIGHLY PLASTIC CLAY OR SILTY CLAY WITH
LENSES, CALCAREOUS NODULES AND FOSSILS IN PARTS.

PI = 33-40

— 40 m

DENSE SAND

NOTE: NO COMMENTS GIVEN ON CONSISTENCY BUT ABOVE DESCRIPTIONS BASED ON SOME
STRENGTHS QUOTED.

NO COMMENT ON STRUCTURE OF CLAY.

NO COMMENT ON WATER TABLE EXCEPT IT MUST HAVE BEEN SHALLOW.

across the base, which has induced a tilt of 1:10.8 or more than 5 degrees from the vertical. While initially there may have been some shear failure of the upper clay horizons, apparently most of the tilt was produced by long-term differential consolidation which is still continuing.

The degree and rate of consolidation of compressible soils under any proposed loads may be crudely estimated by *in situ* tests such as the Dutch probe test or by the standard penetration test (SPT). For normally consolidated clays the compression index (C_c) provides the order of magnitude of the settlement of any foundation; it can be approximately estimated from the liquid limit (LL):

$$C_c = 0.009 \ (LL - 10)$$

(Terzaghi and Peck 1967). Knowing the increase in load on the clay due to the foundation, and the thickness of the stratum, the settlement may be calculated as in Chapter 2. Dense sands and stiff overconsolidated clays will not consolidate nearly as much as loose sands or soft normally consolidated clays, and do not present such severe problems.

More accurate predictions may be obtained by testing undisturbed samples in a consolidometer (p. 270). If the potential settlement thus calculated exceeds what can be tolerated by the structure, its foundations may have to be modified, e.g. by using a raft foundation to spread the applied pressure over a greater area, and to reduce it to a tolerable value (Chapter 4). Alternatively, *in situ* densification methods may be applied, or the loads may be transmitted by piles to deeper and stronger horizons which can support even the greater pressures on a smaller bearing area.

Shrinkage of active soils

Clay soils change their volume on drying out, particularly those containing active clay minerals. Substantial shrinkage can occur when the water table is lowered under drought conditions, which may produce differential settlement in buildings, as there is a greater loss of soil water under their external walls where there is less protection against evaporation. Likewise, deeply rooting vegetation, notably trees, can so desiccate parts of the soil as to cause unsightly cracking in houses. Heaving is the reverse process which accompanies the wetting of active soils above a deep water table (p. 113).

The damage caused by shrinkage settlement totalled many millions of pounds during the drought of 1976 in Britain (Fig. 5.2).

The most severe problems are caused by 2:1 lattice clays, especially of the smectite group which occur in clayey soils of a variety of origins, notably alluvial and marine deposits and clays residual from basic igneous rocks. In tropical and temperate environments the distribution of such soils may often be inferred from geological maps. In arid and semi-arid areas, however, topographic controls are more dominant, and active soils may be encountered in bottom lands on a variety of parent materials (Chapter 3).

Active clays may be inferred from the fissuring and particularly slickensiding apparent in the soil fabric during soil profiling (Chapter 6). The *activity* of the soil, i.e. its susceptibility to volume change on changes in moisture content, is the ratio of the plasticity index to the clay fraction (Fig. 5.3). It is a useful guide to the clay mineralogy of the soil, e.g. pure kaolinite has an activity of about 0.4, pure illite about 0.9, and pure montmorillonite 1.5 to 6.0. An activity of more than 0.5 indicates the likelihood of shrinkage.

The linear shrinkage test (p. 262) provides a further indication of the shrinking and swelling potential of a soil; soils with an LS greater than 8 are susceptible to shrinkage problems.

If the soil is active, the seasonal range of fluctuation of the water table should be determined. Structures may be founded on trench-fill foundations or short-bored piles in the permanently saturated zone. Even where the fluctuations in the water table are likely to be small and shallow foundations are used, care must be taken to avoid local desiccation by trees, wells, boiler-houses, furnaces, or kilns. Water-loving trees with extensive root systems such as poplars, willows, eucalypts, and elms must not be planted at a distance from the building of less than 1.5 times their ultimate height, and adequate insulation or special foundation treatment must be provided for thermal installations.

Sensitivity

Sensitivity denotes the condition of clays that undergo a severe loss in strength after slight disturbance of the soil fabric. The sensitivity S_t is expressed as the ratio of the undrained shear strengths of undisturbed and remoulded soil. All clays exhibit some sensitivity; in normal clays (e.g. London Clay) the sensitivity is less than 4,

FIG. 5.2. Typical cracking produced by shrinkage settlement when active clays were desiccated in the British drought of 1976. (Photograph by J. B. Burland.)

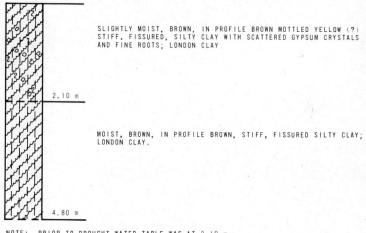

SLIGHTLY MOIST, BROWN, IN PROFILE BROWN MOTTLED YELLOW (?)
STIFF, FISSURED, SILTY CLAY WITH SCATTERED GYPSUM CRYSTALS
AND FINE ROOTS; LONDON CLAY.

2.10 m

MOIST, BROWN, IN PROFILE BROWN, STIFF, FISSURED SILTY CLAY;
LONDON CLAY.

4.80 m

NOTE: PRIOR TO DROUGHT WATER TABLE WAS AT 2.10 m

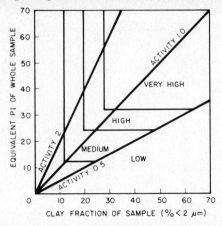

FIG. 5.3. The activity of a soil is a function of its plasticity index and clay content. In routine tests the PI and the clay content are determined from the fraction of the sample $<425\,\mu m$. In order to take into account the behaviour of the whole sample the 'equivalent' PI is adjusted by a factor equal to the proportion of fines on which the tests were conducted. Thus if the PI of a soil containing 75 per cent $<425\,\mu m$ size was 40, the equivalent PI for plotting on the activity chart would be $40 \times 0.75 = 30$. The clay percentage is also adjusted by this factor.

while moderately sensitive clays (e.g. of estuarine and marine origin) have sensitivities of 4 to 8. Some clays of glacial origin are highly sensitive, with sensitivities exceeding 8 and occasionally 100.

Their high sensitivity has resulted from deposition under conditions in which mutually attracted fine particles have been flocculated to produce an open fabric with a very high void ratio; the water held in the voids gives an abnormally high moisture content in excess of the liquid limit. While the clay is undisturbed this is of no great consequence and the shear strength may be relatively high, but when the soil is disturbed, for example by earthquake shock or by pile-driving, the soil fabric collapses and the load is carried by the pore water. This converts the mass to a viscous fluid, which flows even on very gentle slopes until the excess water has been lost and the material reverts to a solid state with lower natural moisture content.

Highly sensitive glacial and glacio-marine clays occur in the valleys of the St. Lawrence and Ottawa rivers in Canada (Leda Clay) and in southern Scandinavia. At a site 35 km northwest of

Oslo, some 200 000 m^3 of material flowed for 1.5 km down a very flat valley to leave a bowl-shaped scar, 200 to 300 m in diameter and 5 to 8 m deep (Fig. 5.4). Bjerrum (1955) reported that the soil was inactive, with a plasticity index of 7 and a clay content of 37 per cent; the natural moisture content was 32 per cent, which was in excess of the liquid limit of 26, and the *in situ* dry density was 13.85 kN m^{-3}. The sensitivity of the material was greater than 10.

A knowledge of the origin of the soil will help to predict sensitivity. Its presence and severity are determined in the field with a shear vane (Appendix B). In the laboratory, the shear strength of undisturbed samples may also be compared with the strength of remoulded samples, but great care must be taken not to disturb the delicate natural fabric of the soil during sampling, which should be done with a single rapid insertion of a thin-walled sampling tube.

It is often difficult and complicated to produce satisfactory foundations, since disturbance of the soil, by pile-driving or by the application of dynamic loads causes a severe loss of strength or even liquefaction, which may threaten the stability of neighbouring structures. In severe cases, the best solution may be to excavate within caissons down to a stable horizon or even to bedrock, or to use raft foundations to reduce foundation pressures to values compatible with the remoulded shear strength of the soil.

PROBLEMS ENCOUNTERED IN AREAS OF DEEP WATER TABLE

Heaving of active soils

Volume changes due to the presence of active clay minerals in the soil are reversible and produce not only shrinkage but also heave when the soil is wetted.

In certain transported and residual soils in arid or semi-arid areas in India, the Middle East, the Americas, Australia, and Africa, where the water table is deep (more than about 5 m but in places more than 100 m) and the soil is dry, the disturbance of natural surface conditions, by irrigation, removal of vegetation, or the emplacement of an impermeable membrane such as road surfaces or buildings, may increase the soil moisture content so that active clay minerals swell and cause differential surface heave of more than 200 mm in severe cases.

FIG. 5.4. Flow in sensitive clays near Oslo.

FIG. 5.5. Typical 'corners-down' cracking pattern produced by dome-shaped heaving distortion in active clays (Afula, Israel). The house is founded at 3 m on augered piles which were not under-reamed to prevent uplift. (After Komornik 1979; photograph by A. B. A. Brink.)

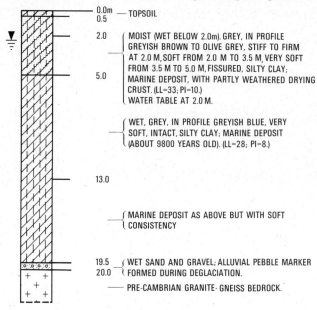

0.0m	TOPSOIL
0.5	
2.0	MOIST (WET BELOW 2.0m). GREY, IN PROFILE GREYISH BROWN TO OLIVE GREY, STIFF TO FIRM AT 2.0 M, SOFT FROM 2.0 M TO 3.5 M, VERY SOFT FROM 3.5 M TO 5.0 M, FISSURED, SILTY CLAY; MARINE DEPOSIT, WITH PARTLY WEATHERED DRYING CRUST. (LL=33; PI=10.) WATER TABLE AT 2.0 M.
5.0	
	WET, GREY, IN PROFILE GREYISH BLUE, VERY SOFT, INTACT, SILTY CLAY; MARINE DEPOSIT (ABOUT 9800 YEARS OLD). (LL=28; PI=8.)
13.0	
	MARINE DEPOSIT AS ABOVE BUT WITH SOFT CONSISTENCY
19.5	WET SAND AND GRAVEL; ALLUVIAL PEBBLE MARKER
20.0	FORMED DURING DEGLACIATION.
	PRE-CAMBRIAN GRANITE- GNEISS BEDROCK.

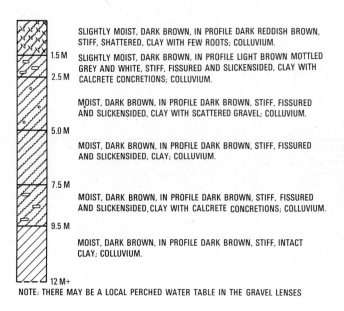

	SLIGHTLY MOIST, DARK BROWN, IN PROFILE DARK REDDISH BROWN, STIFF, SHATTERED, CLAY WITH FEW ROOTS; COLLUVIUM.
1.5 M	SLIGHTLY MOIST, DARK BROWN, IN PROFILE LIGHT BROWN MOTTLED GREY AND WHITE, STIFF, FISSURED AND SLICKENSIDED, CLAY WITH CALCRETE CONCRETIONS; COLLUVIUM.
2.5 M	
	MOIST, DARK BROWN, IN PROFILE DARK BROWN, STIFF, FISSURED AND SLICKENSIDED, CLAY WITH SCATTERED GRAVEL; COLLUVIUM.
5.0 M	
	MOIST, DARK BROWN, IN PROFILE DARK BROWN, STIFF, FISSURED AND SLICKENSIDED, CLAY; COLLUVIUM.
7.5 M	
	MOIST, DARK BROWN, IN PROFILE DARK BROWN, STIFF, FISSURED AND SLICKENSIDED, CLAY WITH CALCRETE CONCRETIONS; COLLUVIUM.
9.5 M	
	MOIST, DARK BROWN, IN PROFILE DARK BROWN, STIFF, INTACT CLAY; COLLUVIUM.
12 M+	

NOTE: THERE MAY BE A LOCAL PERCHED WATER TABLE IN THE GRAVEL LENSES

Lightly loaded structures such as houses are particularly vulnerable; maximum heave takes place under the centre of the structure where the greatest build-up of moisture occurs, and so the effects of heave are manifested in a dome-shaped distortion. The resulting cracks generally widen upwards and are often associated with an apparent 'corners-down' pattern (Fig. 5.5).

Laboratory procedures for identifying active clays have been mentioned above (p. 109). In the field a potentially expansive soil may usually be seen to have a fissured, shattered, or slickensided structure (Chapter 6). An initial assessment of the degree of potential expansiveness (PE) of each soil horizon above the water table is made from the Activity Chart. A better estimate of the maximum potential heave of the soil profile as a whole can be made with Van der Merwe's (1964) formula, namely:

$$\text{total heave} = \sum_{D=1}^{D=n} FD(PE)_D$$

where D is the depth in metres and F is the factor by which heave decreases with depth, obtained from the formula:

$$F = 10^{-D/6.096}$$

Fig. 5.6 may be used to estimate the maximum potential heave from graphical relationships prepared using these two formulae. Heave of the order of 3 per cent of the thickness of a potentially expansive horizon is not uncommon. Differential heave is generally taken as half of the total heave. It is important to assess the actual moisture content of the soil horizons involved, since Van der Merwe's formula estimates the heave on wetting from a desiccated state and therefore gives a maximum value; judgement must be used to correct this value to the natural moisture content.

Brackley (1979) proposes a method for predicting heave which takes into account the natural moisture content of the soil, its *in situ* density or void ratio, and its plasticity index:

$$\text{per cent swell} = \left(5.3 - \frac{147e}{PI} - \log_{10}p\right)(0.525\ PI + 4.08 - 0.85w)$$

where PI is the plasticity index, p is the external load (kPa), e is the void ratio, and w is the moisture content (%). This is integrated over all horizons of the soil profile above the water table to calculate a cumulative heave prediction.

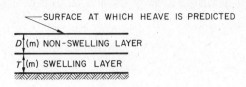

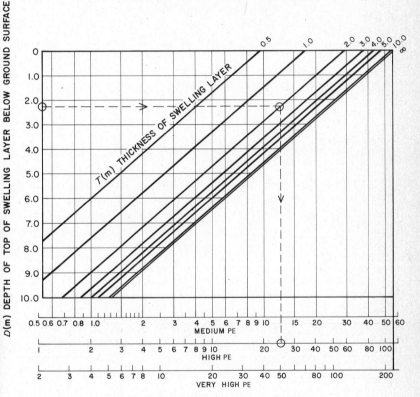

FIG. 5.6. Nomogram for estimating the total potential heave likely to be experienced in active soils. (After Van der Merwe and Savage 1979.)

A more sophisticated method for predicting heave measures volume changes over the full moisture range from dry to saturated on undisturbed samples of the soil by oedometer (consolidometer) (Jennings and Knight 1956). Several foundation and structural solutions have been offered in Chapter 4.

Collapsible fabric

A collapsible fabric may occur in any open-textured silty or sandy soil which has a high void ratio, and yet has a relatively high shear strength in the dry state imparted to it by colloidal or other coatings around the individual grains. When saturated under load, the bridging colloidal material undergoes an instantaneous loss of strength and the grains are forced into a denser state of packing with a reduction in void ratio (Figs. 5.7 and A.1(a) (p. 263)). This may lead to sudden foundation settlement of some magnitude, perhaps up to 10 per cent of the thickness of the potentially collapsible soil horizon.

The condition is common in areas where acid crystalline rocks such as granite or a highly feldspathic sandstone (arkose) have undergone prolonged weathering to produce intensely leached residual soils. In this case the quartz grains are bridged by coatings of kaolinite formed on the decomposition of feldspar (p. 57). The formation of residual soils is particularly rapid in areas with an annual water surplus (i.e. where rainfall exceeds evapo-transpiration): deep residual soils are also found on ancient erosion surfaces which usually form plateaus or benches in the higher parts of the landscape (Chapter 3). For one area, Fig. 5.8 shows the relationship between the remnants of an old erosion surface and areas occupied by residual granitic soils with a collapsible fabric. A collapsible fabric produced by kaolinite bridges is also encountered in sandy colluvial soils, while carbonate bridges are present in some aeolian deposits, such as the loess of China and parts of Central Asia and North America.

The single-storey house of Fig. 5.9 was situated on a colluvially transported, slightly clayey silt-sand, 3 m thick, overlying silty residual shale. It had stood for 40 years without any significant cracking; construction of an external wall, which resulted in local concentration of runoff against the foundations during periods of heavy rain, caused localized collapse settlement of about 150 mm.

A loose or open fabric in a sandy or silty soil and the presence of

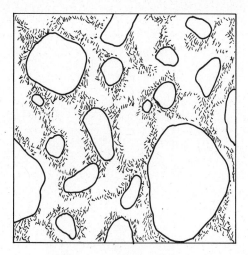

CLAYEY SAND WITH A COLLAPSIBLE
FABRIC (COPIED FROM PHOTOMICRO—
GRAPH X 30)

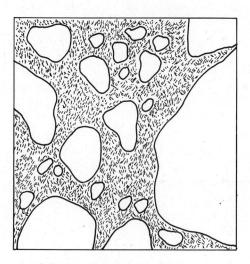

FABRIC OF CLAYEY SAND AFTER
WETTING AND CONSOLIDATION.
(COPIED FROM PHOTOMICROGRAPH X 30)

FIG. 5.7. Soil with a collapsible fabric; based on photomicrographs.

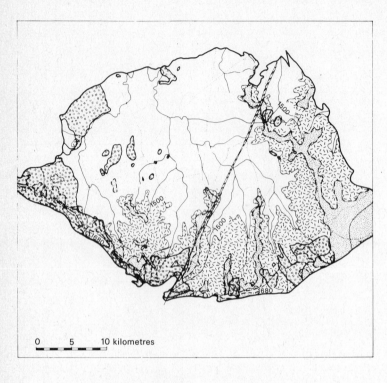

0   5   10 kilometres

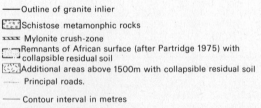

——— Outline of granite inlier

Schistose metamonphic rocks

Mylonite crush-zone

Remnants of African surface (after Partridge 1975) with collapsible residual soil

Additional areas above 1500m with collapsible residual soil

Principal roads.

——— Contour interval in metres

>——— Major drainage pattern

FIG. 5.8. A collapsible fabric often develops to great depth in residual soils beneath old erosion surfaces on granitic rocks.

colloidal coatings around the quartz grains, and clay bridges between them, are easily recognized under a hand-lens, and will alert the observer to the possible existence of this condition. A simple field test is to carve two small cylindrical samples of undisturbed soil as nearly as possible to the same diameter and length, to wet and knead one sample and remould it into a cylinder of the original diameter: an obvious decrease in length when compared with the undisturbed twin sample will confirm a collapsible fabric. A similar reduction in volume may be observed by back-filling a pit: if the soil has a collapsible fabric it will fail to fill the pit completely, in contrast to the increase in bulk experienced with other soils. However, cracking in buildings near a suspect area is the most significant field evidence. Rigid structures tilt towards the area of maximum collapse and unreinforced masonry shows cracking around leaking pipes or gutter outlets where water enters the soil.

Contour maps and aerial photographs may be useful in detecting remnants of old erosion surfaces on which deep residual soil with a collapsible fabric may be present.

There are several laboratory procedures to diagnose the presence of a collapsible fabric. Particle size distribution analysis identifies silty or sandy soils of low clay content; the high void ratios of such collapsible soils impart unusually low dry densities in the range 9 to 16 kN m^{-3}. The amount of collapse settlement which will take place under a specific foundation pressure after saturation of the soil may be predicted by testing twin undisturbed samples in consolidometers, one saturated and the other at natural moisture content (Jennings and Knight 1957). Similar results are obtained with the single oedometer test in which the soil sample is saturated under a particular applied pressure (Jennings and Knight 1975). Fig. 5.10 shows the results of both tests.

There are several engineering solutions to overcome this problem. Where the depth of the collapsible material is less than about 10 m, end-bearing augered piles offer an economic solution. Driven or displacement piles may be used where greater depths are involved. 'Vibroflotation' has also been successfully used to provide safe bearing pressures of up to 400 kPa. A large vibrating cylinder or 'torpedo' is lowered vertically into the sandy soil while water is forced out through jets in its lower end. This compacts the surrounding soils by a combination of vibration and inundation,

FIG. 5.9. The effects of major differential settlement produced by inundation of fine colluvium with a collapsible fabric. (Photograph by T. C. Partridge.)

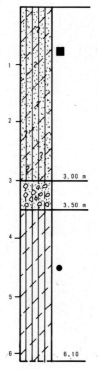

DRY (MOIST, AFTER INUNDATION), DARK REDDISH BROWN, IN PROFILE
REDDISH BROWN, STIFF WHEN DRY (VERY SOFT AFTER INUNDATION),
INTACT, SLIGHTLY CLAYEY SAND SILT WITH ROOTS; COLLUVIUM.
(LL = 22; PI = 7; LS = 4.5 ; CLAY CONTENT = 12 %)

3.00 m

SUBANGULAR, MEDIUM AND FINE VEIN-QUARTZ GRAVELS IN A MATRIX AS ABOVE;
PEBBLE MARKER.

3.50 m

SLIGHTLY MOIST, DARK RED, IN PROFILE DARK RED MOTTLED YELLOW AND
STREAKED BLACK, FIRM, LAMINATED, CLAYEY SILT; RESIDUAL HOSPITAL HILL
SHALE, (LL = 40; PI = 15; LS = 8 ; CLAY CONTENT = 16 %)

6.10

NOTES: 1) SOIL PROFILE RECORDED IN 750 mm DIAMETER HOLE AUGERED BY
HUGHES LDH 60 RIG, NOT TO REFUSAL.

2) WATER TABLE NOT ENCOUNTERED.

3) UNDISTURBED BLOCK SAMPLE TAKEN AT 0.70m; DISTURBED SAMPLE
TAKEN AT 4.50 m.

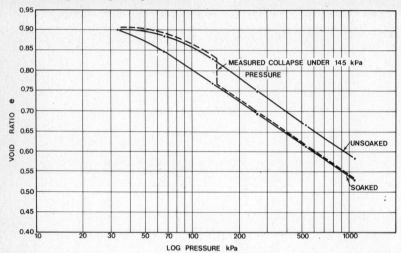

FIG. 5.10. Consolidation curves for residual granite with a collapsible fabric.

and the hole is continually filled with coarse gravel or crushed stone as the torpedo is slowly withdrawn from the soil. The torpedo is typically 600 mm in diameter; after lowering it to 6 m, about 6 m^3 of gravel are required for tamped backfill. The increase in load-bearing capacity results from a combination of soil densification and the gravel columns acting as piles. Dynamic consolidation may also be used (p. 97), though care must be exercised not to affect adjacent structures. Accumulations of water near foundations should be avoided wherever a load must be supported by suspect material.

Sinkholes and subsidences in areas of soluble rock

Sinkholes, dolines, and other manifestations of karst development in areas of soluble rock with a substantial cover of soil are less widespread in their occurrence than most of the other problems discussed in this chapter, but they are probably the most severe and alarming (Brink 1979).

A doline, or compaction subsidence, is a surface depression which appears slowly over a period of years: it may be circular, oval, or linear in plan and may range in size from a few tens of metres to more than a kilometre and may attain a depth of 10 m or more. A sinkhole, on the other hand, appears suddenly, sometimes with catastrophic consequences, as a circular and steep-sided hole in the ground which may attain a diameter of more than 100 m and a depth of more than 50 m.

Common to both of these phenomena is the initial formation of a cavern in the soluble bedrock by the corrosive action of rainwater charged with carbon dioxide (Brink and Partridge 1965). When erosion and downcutting by surface streams have slowly lowered the level of the water table, different sequences of events may form dolines or sinkholes. Compressible materials such as wad, a highly compressible residue of MnO_2 from the solution of dolomite, which have accumulated in the cavern, consolidate slowly on being drained of water and produce a doline (Fig. 5.11). Where the cavern has connected to the overlying soil through a narrow solution-widened joint, or 'slot', which itself is choked with soil, the situation is stable as long as the water table remains above the cavern roof. As soon as the water table drops below this level, rainwater, or more significantly, artificial concentrations of flow in the soil (e.g.

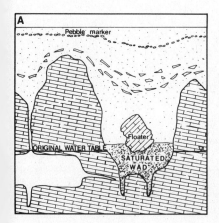
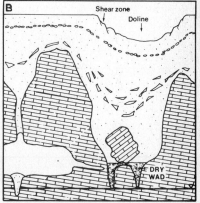

FIG. 5.11. Mechanism of the development of a doline. A shows the equilibrium situation before the lowering of the water table. The palaeo-doline is not apparent at the surface but is indicated by sagging chert rubble and the pebble marker. B is the position after the lowering of the water table. Reactivated doline development becomes apparent when the consolidation of wad leads to surface subsidence. The periphery of a doline is characterized by a shear zone and tension cracks. (After Brink 1979.)

from leaking pipes), may flush the soil from the slot into the air-filled cavern and cause a vault-shaped cavity to develop in the overlying soil. This void is propagated towards the surface by progressive arch collapse which involves the 'peeling' of successive layers from the roof of the void, and the final breakthrough to the surface produces the sinkhole (Fig. 5.12).

The development of these phenomena may be accelerated if the water table is artificially lowered, e.g. by deep mining activities (Fig. 5.13).

If it is known that cavernous rocks underlie a deep soil cover, sophisticated and costly investigation procedures to identify potentially unstable areas are justified. Local variations in bedrock topography, especially the presence of narrow soil-filled slots and of deep sub-surface valleys choked with compressible soils, may be

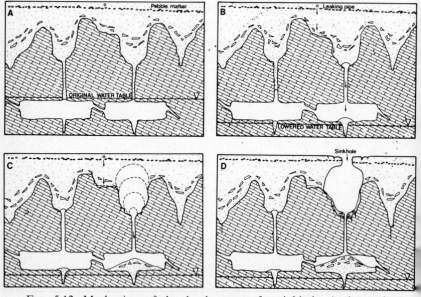

FIG. 5.12. Mechanism of the development of a sinkhole. A shows the equilibrium situation before the lowering of the water table. B is the position after the lowering of the water table. There is active subsurface erosion. The slot is flushed out by a process of headward erosion. C shows the progressive collapse of the roof of the vault, possibly arrested temporarily, by the ferruginized pebble marker. D shows the collapse of the last arch to produce a sinkhole surrounded by concentric tension cracks. (After Brink 1979.)

identified by gravimetric surveys accompanied by exploratory drilling (p. 178). Once the geometry of the buried bedrock surface has been determined in this way, it may be possible to classify and delineate potentially unstable areas in terms of the typical configurations of Figs. 5.11 and 5.12. Such interpretations are of course subject to the limitations of the gravimetric technique.

Where the soil cover is relatively shallow over much of the area, or where it is known that numerous rock pinnacles are separated by deep solution valleys, thermal linescan imagery (p. 325) is often useful to delineate the structural features of the bedrock in detail. Fig. 5.14 compares thermal linescan imagery of a dolomitic terrain with conventional aerial photography of the same area.

Telescopic benchmarks may be installed where there is fear of subsidence or sinkholes near existing structures (Jennings 1966).

FIG. 5.13. A sinkhole which engulfed a three-storey crusher plant in an area of deep residual soil overlying dolomite in the Far West Rand Goldfields of South Africa. (Photograph by courtesy of the *Rand Daily Mail*.)

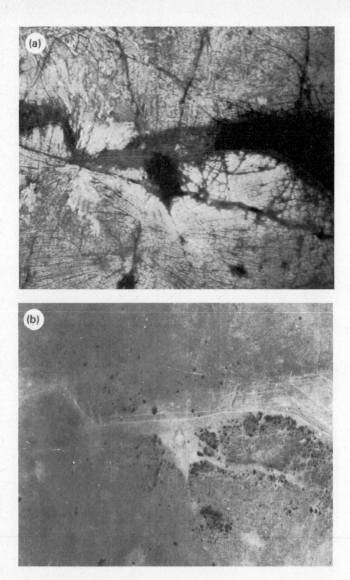

FIG. 5.14. (a) Thermal infrared imagery reveals the fracture pattern in dolomite beneath a thin soil cover. (b) is a conventional panchromatic airphoto of the same area. (Thermal imaging by Spectral Africa Ltd. 1975.)

These penetrate the soil cover and, by means of concentric tubes fixed into the walls of the borehole at different depths, give warnings of differential sub-surface movements which could give rise to dolines or sinkholes (Fig. 5.15).

Engineering development should be avoided where severe problems are expected. If the threat is less severe, the foundations of major structures may be designed to span between pinnacles or high points in the bedrock surface, but care must be taken not to found any part of a structure on isolated rock 'floaters' in the soil. The presence of floaters, and the detailed configuration of the bedrock surface, is best revealed by numerous percussion boreholes; when the boreholes intersect cavities, an indication of their size may be obtained with borehole cameras. When cavernous dolomite and floaters are present, foundation treatment is generally costly and is seldom justified for small structures or housing development. Another possibility is to construct an engineered fill which will span between high points in the bedrock surface in the event of sub-surface movements taking place.

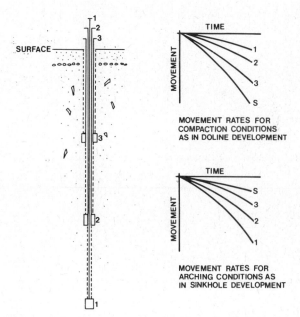

FIG. 5.15. Telescopic benchmarks give early warnings of the development of dolines and sinkholes.

The monitoring of water table levels is an essential prerequisite to any development in areas subject to these problems, in order to notice any lowering of the water table and to take steps, if possible, to prevent it. The control of the discharge of surface water, including leakage from buried pipes, is of the utmost importance in reducing the likelihood of sub-surface soil erosion which could give rise to sinkholes.

PROBLEMS ENCOUNTERED WHERE THE WATER TABLE IS EITHER DEEP OR SHALLOW

Soluble salts in soils

Some salts are aggressive in their effects on buried pipes and on construction materials like concrete and thin bituminous surfacings on roads. The most common of these salts are the sulphates of magnesium, calcium, and iron (e.g. gypsum, $CaSO_4.2H_2O$) and the chlorides of magnesium and sodium (e.g. common salt or halite, NaCl). They occur widely in semi-arid or arid areas, and in humid areas where anaerobic soil conditions favour the presence of

FIG. 5.16. Blisters in a bituminous road surface caused by the concentration of soluble salts from the underlying controlled layers. (Photograph and soil profile by F. Netterberg.)

sulphate-reducing bacteria. In arid areas, high evaporation from the soil surface tends to concentrate the salts near to the soil surface, especially where the underlying bedrock, e.g. marine sediments, contains saline connate water, or in areas subject to periodic flooding. Chlorides are usually most concentrated near the coast where sea breezes carry salt-laden spray inland.

Concrete made from ordinary Portland cement may be corroded and lose its cementing action in a saturated soil with a sulphate content exceeding 0.12 per cent; this is caused by expansion in the cement due to chemical reactions in sulphate solution. Blisters may form in the relatively pervious bituminous surfacings of roads if more than 0.15 per cent of soluble sulphates or more than 0.5 per cent of soluble chlorides are concentrated in the underlying construction materials (Fig. 5.16). This is because dissolved salts are brought towards the surface of the road by capillary action; evaporation through the thin bituminous surfacing exposes concentrations of salts at the points of greatest moisture flux (e.g. at small existing cracks), and these build up to form blisters that lift the surfacing.

Iron pipes corrode in the soil when a soil electrolyte, particularly

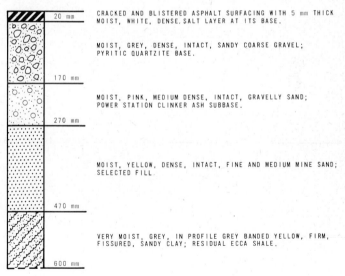

CRACKED AND BLISTERED ASPHALT SURFACING WITH 5 mm THICK MOIST, WHITE, DENSE, SALT LAYER AT ITS BASE.

MOIST, GREY, DENSE, INTACT, SANDY COARSE GRAVEL; PYRITIC QUARTZITE BASE.

MOIST, PINK, MEDIUM DENSE, INTACT, GRAVELLY SAND; POWER STATION CLINKER ASH SUBBASE.

MOIST, YELLOW, DENSE, INTACT, FINE AND MEDIUM MINE SAND; SELECTED FILL.

VERY MOIST, GREY, IN PROFILE GREY BANDED YELLOW, FIRM, FISSURED, SANDY CLAY; RESIDUAL ECCA SHALE.

20 mm
170 mm
270 mm
470 mm
600 mm

NOTE: WATER TABLE NOT ENCOUNTERED.

sulphate, is present, to produce a complete electrical circuit (Corcoran, Jarvis, Mackney, and Stevens 1977); a soil moisture content of more than 20 per cent is necessary for the pore water to function in this way.

Deleterious soluble salts in the soil may sometimes be recognized visually, e.g. the 'desert roses' sometimes formed by intergrowths of gypsum crystals. However, the severity of this problem is usually assessed by chemical analysis.

Electrical conductivity measurements on a saturated paste of the soil fines (Netterberg and Maton 1975) indicate severely aggressive conditions by electrical resistivities of less than 2000 Ωcm in saturated soil, while values greater than 5000 Ωcm indicate non-corrosive conditions (p. 345). This method is relatively insensitive to the less soluble salts such as calcium sulphate. pH itself is not a reliable indicator of the concentration of potentially deleterious salts, but the percentage sulphate (as SO_4^{2-}) can be roughly estimated as 1.5 times the total salt content divided by the pH (Netterberg, Maton, and De Kock 1975). Soils with redox potentials of less than 400 to 430 mV offer favourable environments for the development of sulphate-producing bacteria.

Construction materials with concentrations below the limits referred to above should be preferred; saline soils should be treated with small percentages of high-calcium lime, and low-permeability bituminous road surfacings should be used. Concrete which is to be in contact with saline soils or groundwaters must be made with sulphate-resisting Portland cement or with high alumina cement (Building Research Station 1968), or protected with bituminous or epoxy coatings. Similar coatings may be applied to steel or iron surfaces in contact with corrosive soils.

Dispersion

Sodium-based clay minerals are prone to rapid dispersion or deflocculation in water, and so they are highly susceptible to erosion and to piping when subjected to a hydraulic gradient, for example by seepage through a dam embankment. This has caused the failure of many earth dams; Fig. 5.17 shows the failure of a small earth dam constructed of alluvial clayey silt compacted to 96.5 per cent Proctor density, 2 per cent drier than optimum moisture content. The dam failed on first filling with non-saline flood water.

Soils of several origins are dispersive; particularly suspect are clayey alluvial and gully-wash soils that have been subject to a build-up of exchangeable sodium. In Africa, Mopani trees prefer such saline soils and often indicate their presence on aerial photographs. A columnar or prismatic structure in the subsoil is diagnostic of the dispersive condition.

There are several simple tests for recognizing dispersive soils. No single one is always diagnostic, and so all of the tests should be applied (Sherard and Decker 1977):

(i) Determination of exchangeable sodium percentage (ESP): if exchangeable sodium is more than 7 per cent of the cation exchange capacity the soil is likely to be dispersive; this is the most reliable indicator of potential dispersion.

(ii) A 'crumb test', in which a moist (natural moisture content) pea-sized fragment of the soil is soaked in distilled water to assess the extent to which the particles disperse and form a colloidal cloud.

(iii) The USDA Soil Conservation Service (SCS) 'dispersion test', which compares two hydrometer tests. In one of two glass cylinders a soil–water suspension is prepared by agitating soil with a chemical dispersing agent (such as sodium hexametaphosphate). In the other a soil–water suspension is prepared without either dispersant or agitation. The degree of dispersion is

$$\frac{\text{percentage of particles smaller than 5 } \mu\text{m (treated)}}{\text{percentage of particles smaller than 5 } \mu\text{m (untreated)}}.$$

Clay aggregates in dispersive soils tend to deflocculate without treatment to a greater extent than the aggregates in non-dispersive soils, and the ratio of the two values is therefore larger. Soils with a degree of dispersion exceeding 40 per cent are susceptible to piping erosion, but the critical value may be as low as 25 per cent in some silts, clayey sands, or silty clays of low plasticity.

(iv) The 'pin-hole test', in which a soil sample is remoulded in a cylinder and pierced with a needle of about 2 mm diameter; distilled water is passed through the hole for a specified period, during which the hole is carefully examined for any increase in size, and the discharged water is examined for discoloration due to suspended sediment. The degree of dispersion is assessed from the enlargement of the hole as gauged by the increasing rate of flow and the discoloration of the water. Initially the head of water is 50 mm, but this is increased in stages to about 1000 mm if early results do not indicate dispersive behaviour.

The electrical conductivity of the soil paste provides an additional indicator, not infrequently associated with salinity and dispersive behaviour; conductivity values in excess of 0.5 mS (less

FIG. 5.17. Piping failure in an earth dam constructed of materials containing dispersive clay. The breach was filled with the same silty clay as before, but now treated with gypsum, and gypsum was added to the dam water. The gypsum replaces exchangeable sodium cations, which are loosely held in the clay lattice, with exchangeable calcium cations, which produce a flocculated structure. (Photograph and soil profile by F. von M. Wagener.)

0 m

DRY, GREY, IN PROFILE LIGHT GREYISH BROWN, LOOSE, INTACT, SILTY FINE SAND; AEOLIAN.

0.60 m

SLIGHTLY MOIST, DARK GREY, IN PROFILE BLACKISH GREY, STIFF, SHATTERED (WITH WELL-DEVELOPED COLUMNAR STRUCTURE), SILTY CLAY; ALLUVIUM DERIVED FROM NA-RICH MUDROCK.

1.10

MOIST, BROWN, IN PROFILE BROWN SPECKLED BLACK, STREAKED LIGHT GREY AND MOTTLED WHITE, STIFF, FISSURED, SILTY CLAY WITH CALCRETE NODULES; ALLUVIUM DERIVED FROM NA-RICH MUDROCK.

4.00

MOIST, BROWNISH ORANGE, IN PROFILE BROWN MOTTLED ORANGE, FIRM, FISSURED, SANDY CLAY; ALLUVIUM DERIVED FROM NA-RICH MUDROCK.

4.60

NOTES: 1. PROFILE RECORDED IN BORROW PIT.

2. WATER TABLE NOT ENCOUNTERED.

3. VEGETATION VERY SPARSE.

4. ONLY MATERIALS FROM 0.60 m TO 4.00 m USED FOR CLAY CORE OF DAM.

TEST DATA: 0.60 – 1.10; pH   5.9
CONDUCTIVITY 390 ohm cm
ESP   25 %
LL   34
PI   16
LS   9
SILT   39 %
CLAY   48 %

1.10 – 4.00; pH   8.2
CONDUCTIVITY 300 ohm cm
ESP   22 %
LL   50
PI   30
LS   12
SILT   25 %
CLAY   68 %

than 2000 Ω cm resistance) may be associated with dispersive characteristics.

Where dispersive soils must be used in the construction of an embankment dam, the tendency for the soil aggregates to deflocculate may be reduced by adding gypsum or lime; the former is preferred. Alternatively an impermeable membrane, e.g. of plastic, may be installed on the upstream face of the embankment, or the embankment material may be injected with chemical grout to reduce its permeability and seal fissures which may be potential conduits and give rise to piping. Another solution is to construct a flexible clay core by means of carefully controlled compaction at a moisture content *wet* of the Proctor optimum value. However, all of these solutions are costly and it is usually more economic to find more suitable construction materials.

Special problems in arctic regions

In areas where the average annual air temperature is less than about − 2 °C, water in the pore spaces of the soil freezes to a considerable depth below the surface. This phenomenon, known as *permafrost*, is particularly prevalent in northern Canada and the USSR where freezing may reach depths of up to 400 m in places. Fig. 5.18 shows the approximate limits of continuous and discontinuous permafrost in the northern hemisphere.

Permafrost raises problems only when water is present in the soil; the effects are most severe in moist silts and clays which have a high porosity and a low permeability. Dry soils are not affected, but even a low moisture content can result in the development of the ice lenses (Fig. 5.19). Linell and Tedrow (1981) have discussed the problems of classifying and mapping permafrost.

Since water undergoes a 9 per cent increase in volume on freezing, frozen soils are subject to the reverse process on thawing and the consequent settlement may cause severe damage to structures in areas of discontinuous permafrost. Thawing may result from the higher temperature below buildings in areas of continuous permafrost. In either case soup-like conditions may result, especially where frost during the deposition of fine sediments had originally entrapped water; on thawing, the water cannot drain from these oversaturated zones owing to the impermeability of the underlying frozen ground. Where the terrain slopes, even gently, the saturated soil may flow downslope at rates from a few metres per

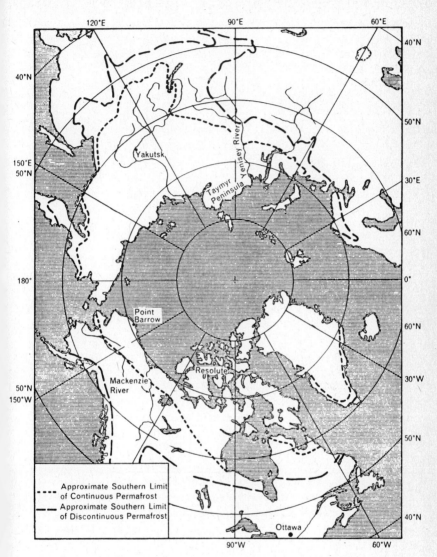

FIG. 5.18. Distribution of permafrost in the northern hemisphere. (After Brown and Johnston 1964.)

FIG. 5.19. Ice lenses developed in an arctic soil profile. (Photograph by R. J. E. Brown.)

year to a few metres per day. This *solifluction* can be extremely damaging to engineering structures (Brown 1970).

Organic matter accumulates on the surface of almost all sub-arctic soils, and is prevented from decaying by the lack of bacteria at these low temperatures. The resulting organic mantle, known as *muskeg*, forms a very old ground cover, usually up to one metre thick and very rarely exceeding ten metres. The fibrous and very porous nature of this material, in which moisture contents of up to 1500 per cent are not uncommon, causes it to act as a protective cover over frozen soils owing to its absorption of solar radiation and the resulting evaporation of thawed-out moisture from the frozen muskeg layer. If the muskeg layer is not disturbed, melting of the ice rarely extends to the underlying soil.

The areas of discontinuous permafrost shown in Fig. 5.18 are readily identifiable in the field and on aerial photographs from their unique 'beaded' drainage pattern and their thermokarst topography, which includes thaw-sinks, pingos (or frost mounds), and solifluction lobes.

A common approach in attempting to solve the engineering problems associated with permafrost involves the maintenance of frozen conditions by preserving the muskeg cover and insulating the structure from the frozen soil with 'brush' (scrub trees and branches) onto which a suitable dry, freely draining material such as sand or gravel is dumped. This solution is applicable in the case of simple structures such as small houses, roads, and airfields, but heavy structures are usually founded on piles steamed into the permafrost or dropped into pre-drilled holes; the space around the pile is filled with slurry to freeze the pile into the soil. Floors of piled buildings are located about one metre above ground level and are well insulated: winter winds blowing through the gap provide ventilation to counteract any thawing in the upper soil that takes place in summer. Water-bearing services are placed in insulated boxes (utilidors) which are treated in the same way by emplacement above the ground on gravel fill or on piles.

In road construction, care is taken to avoid all areas of solifluction; owing to the impermeable nature of the subsoil, the provision of adequate drainage is vital. Failure to provide drainage in disturbed areas may complicate construction through the occurrence of intensified frost-action during winter and accelerated thaw during summer. Where possible, cuts are avoided and the road bed is placed on an insulating fill to prevent thawing.

In dam construction, the frozen ground is either excavated down to bedrock to provide an impermeable cutoff or the frozen condition is maintained, where necessary, by artificial refrigeration (Brown and Johnston 1964).

Extremely variable soils

Several kinds of soil are characterized by extreme variations in texture and in engineering properties over short distances, e.g. till and some soils residual from igneous rocks and from metamorphic rocks. These rapid variations may include hard layers in juxtaposition with soft compressible materials, or large boulders in a fine matrix, and pose problems not only in site investigation but also in foundation design. Legget (1979) states that 'of all the problems encountered with soil in civil engineering works ... the varying characteristics of till (glacial till and "boulder clay" being common but somewhat inaccurate alternative names) have probably created more trouble than any other cause, possibly than all other soil problems combined, at least in the northern hemisphere'. The same may be said of soils residual from ancient igneous and metamorphic rocks in the interior continental masses of Africa, South America, Asia and Australia; the stark variability in the deep residual soils of Hong Kong has been well documented by Lumb (1965).

6. Recording and interpreting the soil profile

An accurate description of the soil profile is a basic requirement in site investigation for any engineering structure. The profile can be properly described only if the soil is examined *in situ*, in its natural state. To provide for this, trial holes should be dug to the depth of stress influence. A useful rule of thumb gives this depth as $1.5B$ where B is the width, or minimum plan dimension, of the structure. Foundation design for pavements (roads or runways), pipelines, and canals generally requires a shallow depth of investigation, less than about 3 m, and the same applies to prospecting pits for materials to be used in various earthworks such as embankments, road prisms and earth dams; hand-dug pits will generally be economical for these purposes. Foundation design for houses and other light structures usually requires trial holes to a depth not exceeding 6 m, and these may be dug with light machines such as back-actors. Major structures require the soil profile to be recorded right down to bedrock or to depths of up to 50 m or more: trial holes for this purpose are best dug with a large diameter (600 to 1000 mm) auger machine (Fig. 6.1). The profile should be examined soon after completion.

If the trial hole is too deep for the observer to descend by ladder, as is usually the case with machine augered holes, he is lowered in a bosun's chair suspended from a light hand-winch. He should wear a hard hat and be strapped into a safety harness attached to a rope which is manipulated from above. A code of practice (e.g. South African Institution of Civil Engineers 1980) must be strictly applied to ensure the safety of the observer, and emphasis is laid on the competence and experience of the man in charge of the operation. Above the water table there is little danger of the sides of the hole collapsing unless the soil is totally non-cohesive, but the observer should descend below the water table only with great caution and should keep looking upwards for any signs of pressure-bursting on the sides of the hole.

Trial holes dug by back-actor are, as a rule, less stable than circular holes, and require shoring below 1.5 m, particularly if the hole extends below the water table. Where it is unsafe to descend a trial hole, it will be necessary to describe the soil profile from samples retrieved from small-diameter boreholes, as discussed at the end of this chapter. However, these procedures are more costly and less satisfactory than direct observation, both visual and tactile, within a trial hole.

Before descending the trial hole, the observer secures the zero end of a measuring tape at the surface by means of a long nail driven into the soil on the northern edge of the hole: he may then orientate himself as he descends, carrying the tape with him and allowing it to unroll as he proceeds. He should also carry a geological pick, a small bottle of water and a pocket-knife. It may be necessary to use an electric light in deep holes, but far better illumination is achieved by an assistant at the surface reflecting sunlight down the hole with a convex mirror. In the confined space of an augered hole it is difficult for the observer to write his observations in a notebook: a small tape-recorder is more useful. Best of all is an assistant at the surface, booking the verbal descriptions and checking that each soil horizon has been described in terms of the six descriptors—

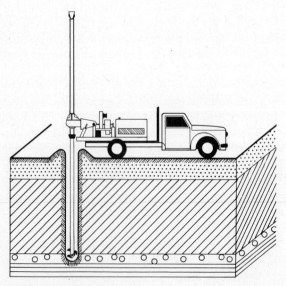

FIG. 6.1. Deep trial holes for soil profiling are best excavated with a large-diameter truck-mounted earth auger.

moisture condition, colour, consistency, structure, soil texture, and origin, or MCCSSO for short. The descriptors are mainly qualitative but it is possible to assign numerical values to some empirically. Each descriptor is related to a soil property of engineering significance and must therefore be assessed in a systematic way (Jennings, Brink, and Williams 1973; Jennings and Brink 1978).

MOISTURE CONDITION

The moisture condition is assessed as dry, slightly moist, moist, very moist, or wet. The moisture condition is a necessary adjunct to the assessment of consistency. At the time of examination the soil may be dry, but the engineer must think what the final condition may have become many years after completion of the structure. For example the initial state of a clayey soil may be dry, with a corresponding very stiff consistency. But, after a building has shielded the surface from evaporation for ten years, the soil may have become very moist and the consistency reduced to soft; its strength will thus have decreased substantially and, if the clay is active, heave will have taken place in the building.

The empirical assessment of moisture content from the description of the moisture condition will depend on the texture of the soil. Sand with a moisture content of 20 per cent will be observed to be wet, but clay at the same moisture content would be described as slightly moist. Regardless of the texture, however, the correct assessment of the moisture condition provides a useful indication of the amount of water required for compaction: dry and slightly moist soils will require the addition of water to attain their optimum moisture content, wet soils (from below the water table) and very moist soils will need to be dried before compaction, while moist soils will be near their optimum moisture content.

COLOUR

The colour of a soil is perhaps its most obvious attribute for correlating a particular horizon with the same one encountered in other holes on the site. It may also serve as the most obvious identification feature in passing instructions to the site foreman; e.g. 'excavate down to the yellow material and stockpile the brown topsoil and the red subsoil separately'. Soil colour may change somewhat with changes in the moisture content from one hole to the next, and so it should be described at a standard moisture condition, the simplest being completely wet. For this, water is added to a few grams of the soil in the palm of the hand and worked into a thick slurry, the colour of which is identified from a standard colour chart. As this will often differ from the colour of the undisturbed soil at its natural moisture content, the colour 'in profile' should also be described. Thus a full description of colour may read 'dark red, in profile brownish red mottled black and blotched yellow'—'blotches' referring to larger patches than 'mottles'.

Besides serving as a recognition feature, colour often indicates chemical processes, mainly those associated with iron compounds. In temperate environments gleying, or the reduction of ferric compounds and the chelation of ferrous ions, is a common result of anaerobic conditions in an horizon that also contains organic matter. It is often diagnostic of the ground water table and its capillary fringe, or of a seasonal water table perched above an impermeable horizon. In the absence of other indications these may be distinguished on:

(i) waterlogging by ground water table—most gleyed *within* structural units,† more mottled outside;
(ii) waterlogging by seasonal perched water table—more gleyed on *outside* of structural units than inside.

These features may not develop, or may not be apparent, in very deep waterlogged horizons, or in vlei or topomorphic vertisol soils, or in peaty soils.

In tropical environments laterization, which may be regarded as an extreme form of gleying, produces black, yellow, or red mottling or blotching which are also associated with changes in the clay mineralogy. The development of 'ironstone' or 'murram' at sites above the bottom of the catena may be associated with changes in vegetation, or changes in water regime, and not necessarily with waterlogging: they are not always associated with gley features. Soils residual from basic igneous rocks commonly show other colour segregations such as white 'speckles' (kaolinized feldspars) and black 'streaks' (relict joints stained with iron or manganese compounds).

† By structural unit is meant an intact lump of soil bounded by natural discontinuities such as fissures, i.e. the *ped* of the pedologist.

Whether or not a standard colour chart (e.g. the Munsell chart) is used, words used to describe colour should be simple, such as dark brown, light olive, light reddish brown, dark yellowish orange, and so on. In his valuable little book on the writing of geotechnical reports, Palmer (1957) has this to say:

> Elegance is all very well, but a fairly golden rule whenever we think we have written something particularly good and elegant, in fact when we are thoroughly delighted with the turn of phrase, is 'strike it out'. Dr. Cooling advises 'when in doubt, wheel it out'. An example of such elegance worthy of note is the engineer's description 'sky-blue clay streaked with claret'. This was wheeled out!

CONSISTENCY

Consistency is a measure of the hardness or denseness of a soil and can be assessed by simple field tests such as the resistance to penetration of the pointed end of a geological pick or resistance to moulding in the fingers. The consistency of cohesive and non-cohesive soils is described using five terms for each as listed in Tables 6.1 and 6.2. These terms have long traditional usage dating from the last century, when they were used to assess 'safe bearing capacity'. Providing no changes take place in the soil as a result of construction, the ranges of unconfined compressive strength and of dry density corresponding to these terms may be used to provide conservative estimates of the strength of an *individual* soil horizon. However, the assessment of safe bearing pressure at the chosen depth of founding must be based on the soil profile as a whole. The engineer thus takes into account such crucial factors as the depth and possible fluctuation of the water table, the presence or absence of harder or softer horizons below the founding depth but within the zone of stress influence, the presence or absence of joints or other planes of weakness in the soil, and so on. Consistency also gives an empirical guide to the compressibility of each soil horizon: a stiff soil will consolidate under load less than a soft one.

In assessing consistency using the sharp end of a geological pick to penetrate the undisturbed soil, the shearing resistance of the intact material reflects its 'peak strength'. Confirmation of this assessment is sought from the initial distortion of an intact lump of the soil on being remoulded in the fingers. Further remoulding, without the addition of water, may produce a considerable reduction in resistance to deformation, i.e. a reduced shear strength; but

TABLE 6.1

Consistencies of non-cohesive soils. (After Jennings, Brink, and Williams 1973)

	Gravels and clean sands—generally free-draining (ϕ materials)	Typical dry density (kN m^{-3})
Very loose	Crumbles very easily when scraped with geological pick	less than 14.50
Loose	Small resistance to penetration by sharp end of geological pick	14.50 to 16.00
Medium dense	Considerable resistance to penetration by sharp end of geological pick	16.00 to 17.50
Dense	Very high resistance to penetration of sharp end of geological pick—requires many blows of pick for excavation	17.50 to 19.25
Very dense	High resistance to repeated blows of geological pick—requires power tools for excavation	more than 19.25

TABLE 6.2

Consistencies of cohesive soils. (After Jennings, Brink, and Williams 1973)

	Silts and clays and combinations of silts and clays with sand, generally slow-draining ($\phi = 0$ materials)	Unconfined compressive strength (kPa)
Very soft	Pick head can easily be pushed in to the shaft of handle. Easily moulded by fingers	<40
Soft	Easily penetrated by thumb; sharp end of pick can be pushed in 30–40 mm; moulded by fingers with some pressure	40–80
Firm	Indented by thumb with effort; sharp end of pick can be pushed in up to 10 mm; very difficult to mould with fingers. Can just be penetrated with an ordinary hand spade	80–160
Stiff	Penetrated by thumb-nail; slight indentation produced by pushing pick point into soil; cannot be moulded by fingers. Requires hand pick for excavation	160–320
Very stiff	Indented by thumb-nail with difficulty; slight indentation produced by blow of pick point. Requires power tools for excavation	320–700

this may still not be the minimum, or 'residual strength', associated with shearing movement along a natural slickensided joint in the soil mass. If clearly noticeable, reduction of strength on remoulding should be noted, as this may be an indication of 'sensitivity' (p. 109)—but it is the consistency of the *in situ* soil that must always be assessed.

Appendix E presents a guide to the assessment of rock consistency. The description of rock encountered at the depth of refusal to a machine auger, or encountered as interbedded layers within the soil, should always be included in the record of the soil profile. It will be seen from Table 6.2 that the upper limit of the unconfined compressive strength of a very stiff soil, and from Appendix E that the lower limit for very soft rock, is about 700 kPa.

The safe bearing pressure for a particular structure founded at a particular depth in the soil profile may be conservatively found from consistency assessments, taking into account the fact that these may reduce if the soil wets. In many cases no further investigation by laboratory or *in situ* testing need be undertaken since foundations may be economically designed to a sufficient size to carry the applied loads and, even if the soil were stronger, it would often not be convenient to reduce the size of the foundations. From its consistency the engineer may also assess the undrained shear strength of the soil—about half the unconfined compressive strength of a wet soil—and from this he is able to calculate, for example, the friction on the base of a retaining wall or the stability of a slope.

SOIL STRUCTURE

By soil structure is meant the presence or absence of fissures or other planes of weakness in the soil. If there are no 'cracks' in the soil as is commonly the case in sands, its structure is described as *intact*. It is *fissured* if it contains closed joints, *shattered* if it contains open fissures filled with air or water, *slickensided* if it contains highly polished shear planes, and *micro-shattered* if it crumbles to the size of coarse sand or fine gravel particles which will break down to clayey or silty material when rubbed with water on the palm of the hand. A slickensided, shattered, micro-shattered, and sometimes a fissured structure will lead the observer to suspect the presence of active clays. In some residual soils which are not active, fissures or joint planes may have been inherited from the parent rock. Residual soils derived from sedimentary rocks may be *stratified* (parallel bedding planes), *laminated* (in layers less than about 20 mm thick), or *varved* (alternating silty and clayey layers of different colour in a glacial deposit). Residual soils derived from metamorphic rocks will often be *foliated*.

Fissures, and particularly slickensides, affect the overall strength of a soil mass, especially in cut slopes and other types of excavation, and they can also aggravate the effects of water. It is sometimes important therefore to record the orientation of any such planes of weakness; e.g. where bedding planes dip towards the face of an excavation they may cause instability of the face, whereas the opposite face, into which the planes dip, may be perfectly stable. Fissured soils in the sides of excavations tend to behave like cohesionless materials with failure taking place along the fissures rather than along circular or other non-planar sliding surfaces.

Pedologists rely greatly on soil structure to predict soil strength, permeability, and water retention, but they are interested in shallower depths than engineers.

SOIL TEXTURE

The soil texture of each horizon of the profile is described on the basis of the grain size of the individual particles (Table 6.3). The basic textural classes are clay, silt, sand, and gravel, but most natural soils are combinations of two or more of these. In describing such combinations the noun is used to denote the predominant class, e.g. a silty clay is a clay with some silt, but a clay–silt contains roughly equal proportions of each. The simple identification tests given in Table 6.3 usually suffice to distinguish between the classes but grading analyses lend confidence to the assessment of textural combinations. The inexperienced soil profiler usually has most difficulty in distinguishing between clay and silt, and here the dilatancy test is useful: a cake of wet, remoulded soil is shaken between cupped hands until a film of water appears on its surface—if the cake is then squeezed between forefinger and thumb the film of water will retract into the cake if it is a silt. This is because silt is a dilatant material, i.e. it increases in volume as it is sheared. If the hands are allowed to dry after completion of the test,

TABLE 6.3

Particle size classes commonly used in engineering (the Massachusetts Institute of Technology classification)

Grain size (mm)	Classification	Individual particles visible using	Mineralogical composition	Identification test
Less than 0.002	Clay	Electron microscope	Secondary minerals (clay minerals and Fe-oxides)	Feels sticky. Soils hands. Shiny when wet
0.002 to 0.06	Silt	Microscope	Primary and secondary minerals	Chalky feel on teeth. When dry rubs off hands. Dilatant
0.06 to 0.2	Fine sand	Hand lens	Primary minerals (mainly quartz)	Gritty feel on teeth
0.2 to 0.6	Medium sand			
0.6 to 2.0	Coarse sand			
2.0 to 6.0	Fine gravel	Naked eye	Rocks (sometimes vein quartz)	Observed with naked eye
6.0 to 20.0	Medium gravel			
20.0 to 60.0	Coarse gravel			
60.0 to 200.0	Cobbles			
More than 200	Boulders		Rocks	

silt may be easily brushed off them but clay will stick. The feel of the soil between the teeth is also a good aid to identification.

The shape of gravels and boulders should be described as this often aids the interpretation of the origin of the soil. Terms used are:

(i) well-rounded (nearly spherical);
(ii) rounded (tending to oval shape);
(iii) sub-rounded (all corners rounded off);
(iv) sub-angular (corners slightly bevelled);
(v) angular (corners sharp or irregular).

The importance of texture lies in the drainage characteristics of the soil. Gravels and sands drain rapidly while intact clays drain very slowly under load. The rate at which the soil drains under load, and hence the rate at which consolidation takes place, governs its stress–strain behaviour. While the shear strength of gravels is controlled entirely by the total applied load, normally the shear strength of clays will not be affected immediately by the application of a load. Except with pure sands or gravels, engineers tend to consider shear strengths in terms of the undrained, or $\phi = 0$, condition of the saturated soil. In most cases this gives conservative assessments of shear strength, and it is on these that the interpretations of consistency classes in Table 6.2 are based.

ORIGIN

The engineering significance of the origin of different horizons of the soil profile cannot be overemphasized. Correct identification of origin is the first step towards making meaningful predictions of the likely engineering behaviour of a soil. While it is usually quite easy to identify the origin of residual soils, particularly where their upper limit is demarcated by the pebble marker and the bedrock is visible at the bottom of the trial hole, the genesis of transported soils is sometimes more difficult to determine, particularly where the fabric has been transformed by pedogenesis.

Residual soils

A knowledge of the local stratigraphic succession is basic to the correct identification of the origin of the residual soils on any site. Primary rock structures inherited from the parent rock, e.g. bedding planes, joints, foliations, etc., are also helpful. It is

TABLE 6.4

Engineering problems associated with residual soils

Residual soil category	Problems to be expected
Residual acid crystalline rock	Collapsible fabric; high erodibility; presence of core-stones in the soil; pseudokarst phenomena (Brink 1979)
Residual basic and intermediate igneous rock	Active clays; high compressibility; low shear strength
Residual calcareous rock	Sinkhole development; doline development; susceptibility to frost wedging
Residual argillaceous rock	Active clays; slope instability
Residual arenaceous rock	Problems not common but collapsible fabric may be present in soils derived from arkose or highly feldspathic sandstone

sometimes possible to identify primary and secondary minerals which characterize the mineralogical composition of the parent rock. Residual soils derived from acid crystalline rocks such as granite or gneiss will contain primary quartz grains and mica flakes and secondary kaolinite formed by the decomposition of feldspars. Those derived from basic or intermediate igneous rocks, such as dolerite or diabase, will contain no quartz (other than vein quartz in the form of gravel) and will consist of clay or silt; any sand particles will consist of primary feldspars or ferromagnesian minerals. Amygdales in a residual soil demonstrate its derivation from a volcanic parent rock such as andesite or basalt. The identification of bedrock at the depth of refusal to a mechanical auger is often a direct indication of the origin of the overlying residual soils.

The engineering problems likely to be associated with the five broad categories of residual soil listed in Table 3.6 (p. 59) are given in Table 6.4.

Transported soils

There is often a clear relationship between the landform and the origin of transported soil horizons, though it should be remembered that the transported zone of any soil profile may contain horizons of different origin (Fig. 3.3). Consequently the identification of the landform, either in the field or by the interpretation of aerial photographs, is often a great help in identifying the origin of the associated transported soils. But clues must also be sought in the trial hole, for example from the texture of the soil (Table 3.7, p. 60), the degree of rounding or angularity of the gravel fragments (particularly in the pebble marker), and even the presence of stone artefacts or other man-made objects.

TABLE 6.5
Engineering problems associated with transported soils

Transported soil type	Problems to be expected
Biotically reworked soil	Collapsible fabric; high compressibility
Slide debris	Slope instability; high permeability
Talus (coarse colluvium)	Slope instability
Pediment deposit (fine colluvium)	Collapsible fabric (sands); active clays (clayey soils); high compressibility (silty soils); dispersive characteristics (clayey soils)
Pyroclastic deposit	Low strength; high compressibility; slope instability
Gullywash	Dispersive characteristics; active clays; high compressibility; high erodibility
Lacustrine deposit	High compressibility; active clays; high soluble salt content
Till	Extreme variability
Aeolian deposit	Collapsible fabric; high compressibility; mobility (dunes)
Alluvium	Active clays; subject to flooding
Estuarine and deltaic deposits	High sensitivity; high compressibility; quicksand; high soluble salt content
Littoral and marine deposits	Collapsible fabric; instability of dredged excavations; high soluble salt content

Engineering problems which are likely to be associated with the twelve types of transported soil listed in Table 3.7 are given in Table 6.5.

Pedogenic materials

Pedogenic processes acting on either residual or transported soils may radically alter their engineering properties, especially when pedocretes have formed. The terms ferricrete, calcrete, and silcrete refer to soil horizons which have become strongly cemented or replaced by iron oxide, calcium carbonate, and silica, respectively. Depending on their stage of development these materials may appear in various forms; e.g. in the case of calcretes it is possible to distinguish between calcified soils, powder calcretes, nodular calcretes, honeycomb calcretes, hardpan calcretes, and boulder calcretes, all of which have been shown to possess significantly different ranges of engineering properties (Netterberg 1971).

Scattered discrete nodules or concretions in a soil matrix should be described as inclusions, while the more highly developed stages of nodular and hardpan pedocretes should be described as soil profile horizons. In the former case a suitable description might be 'with scattered black ferricrete concretions up to 50 mm in diameter'. In the latter case a hardpan pedocrete might be described as 'moist, reddish brown, in profile brown mottled red and black, intact, soft rock; hardpan ferricrete'.

OTHER DATA RECORDED ON THE PROFILE SHEET

In addition to the six descriptors for each horizon, the following must also be recorded:

(i) the presence of any inclusions in a horizon, e.g. roots, termite channels, gypsum crystals, etc.; these can be important indicators of local conditions, e.g. termites cannot live below the water table, while gypsum is a sign of desiccation;

(ii) the depth of the water table or, if it is not encountered, a statement to this effect;

(iii) the depth at which soil samples have been taken;

(iv) the material at the bottom of the trial hole and the reason for stopping at this depth, e.g. refusal of mechanical auger;

(v) the dates on which the trial hole was excavated and profiled;

(vi) the initials of the observer and the booker, and the organization they represent.

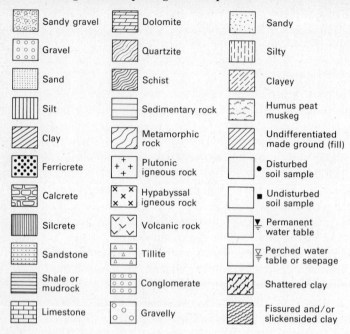

FIG. 6.2. Suggested symbols for use in drawing up soil profile records.

TABLE 6.6

Limitations on profiling from soil samples recovered by the various techniques described in Appendix B, listed in general order of increasing soil disturbance and decreasing cost

Type of sample	Descriptors which can be assessed					
	M	C	C	S	S	O
Continuous thin-walled tube sample	✓	✓	*✓	✓	✓	✓
Cores obtained by rotary drilling		✓	*?	✓	✓	✓
Small-diameter auger spoil	✓	✓	*?		✓	?
Spoon or channel sample	✓	✓	*		✓	?
Percussion drill cuttings	✓	✓	*			?
Wash-boring sediment		✓	*?			?

* Consistency may be judged from the rate of penetration of the sampler.

The textural classes of each soil horizon are depicted on the profile sheet using standard hatchings; those suggested in Fig. 6.2 have been found practical.† A good graphical picture emerges from the hatchings when the engineer lays out a number of profile sheets on a table to produce various cross-sections through the subsoils of the site—this is the starting point in his foundation design.

SOIL PROFILING FROM SAMPLES

When for any reason it is not possible for the observer to descend a trial hole to record the soil profile—the most common reason being the presence of a shallow water table—he is forced to do the best he can from soil samples retrieved by a variety of sampling techniques (Appendix B). Most of these techniques disturb the natural fabric of the soil and interfere with its natural moisture content, and impose limits on the accuracy with which the six descriptors (MCCSSO) can be assessed (Table 6.6).

† There are examples of profiles described in this way on pages 107, 123, 135, and 163.

7. Investigation procedures for major structures on sites of limited area

The site investigation procedures considered in this chapter are used to gather essential data for foundation and materials design in practically all engineering projects, and particularly for heavy structures (p. 94). The engineer designing the foundations of such major structures needs a three-dimensional picture of the distribution and properties of all the different kinds of soil materials beneath the site to relate the proposed structure to the soil (or rock) which will provide its support, as discussed in Chapter 4.

PRELIMINARY INVESTIGATION

The preliminary investigation requires the soil profile to be recorded at positions all over the construction site. The level of sophistication of this stage of the work is perhaps comparable to the type of investigation conducted by a competent engineer of the last century who, without the branch of knowledge known today as *geotechnical engineering* (or soil mechanics), still came to very reasonable conclusions and built works which are impressive even by today's standards. This he did simply by the application of the engineering principles known to him, together with sound engineering judgement (Jennings and Brink 1978).

The following five steps constitute the framework for preliminary investigation:

  (i) understand the nature of the proposed structure and any particular requirements of its foundations;
 (ii) understand the geology and geomorphology of the site;
(iii) conduct surface inspections and enquiries to gather information on the history of the site, its vegetation, and farming or other local practices, and the presence of any signs of distress in existing buildings or other structures nearby which may have resulted from subsoil conditions;
 (iv) record soil profiles throughout the site;
  (v) take disturbed soil samples of any soil horizons of particular concern or interest for indicator testing.

Nature of the structure

The main soil data required when designing the foundations of a major structure to be founded in the soil concern the stress–strain behaviour from which its compressibility and strength can be determined. In seeking depths of founding at which the relevant parameters will be acceptable, the engineer must also consider the differential movements which can be tolerated by the different parts of his structure (Table 4.1) and any effects the structure may have on the soil. For example, in planning a steelworks plant or a glass-making factory, which incorporate smelting, machining, and warehouse storage of heavy products, the following requirements will be vital:

(i) unless thoroughly insulated, the furnace will tend to 'bake' the soil, which could result in large volume changes;
(ii) sensitive machinery may not be able to tolerate any differential movements at all;
(iii) floor loads in the warehouse will be high and may vary considerably from one part of the floor to another and from time to time;
(iv) while the external cladding of the building that houses the factory may be able to tolerate very considerable differential movements, any gantries supported on rails spanning between columns will have only limited tolerance for such movements.

Other structures have other special requirements for stability, limits of deformations, risks of failure, and so on. These requirements must continually be at the back of the engineer's mind as he tailors the site investigation programme to provide the relevant data for design solutions.

Geology and geomorphology of the site

Any available information on the geomorphological history of the site, the geological succession underlying it, and the drainage or runoff from its surface should be assembled at the outset.

The stage which has been reached by the operative geomorphological cycle can be a valuable guide to probable soil conditions; e.g. protracted weathering beneath an ancient erosion surface may have resulted in deep and highly leached residual soils (Partridge 1975).

The design, construction, and maintenance of a structure depend greatly on whether the land surface on which it is built has been sculptured by erosive or depositional agencies, or by slides in the slopes of the hillsides. Slope failures usually take place only in areas of high relief and during periods of unusually heavy rainfall; their effects are potentially most disastrous where the hillsides have a deep mantle of residual soil. Dam basins in such environments are particularly vulnerable as slope failure may be accelerated in the artificially saturated residual soils, which may result in sudden overtopping or even failure of the dam wall. Cuts and fills for roads and railways in areas of 'slide topography' require special design precautions as their presence inevitably leads to an unnatural build-up or release of water pressures which may also accelerate slope failures.

Geological maps and reports are likely to be more useful than soil maps and airphotos during the preliminary stages of site investigation. In a hitherto undeveloped area, however, aerial photographs may be all that are available: with the aid of a terrain classification (Chapter 11) and a data bank (Chapter 12), they may provide an essential means of access to relevant data generated elsewhere in areas of similar geology, climate and geomorphological history.

Site inspections and local enquiries

Site inspections should be conducted in a systematic way with a check list of the things that should be noted: e.g. rock outcrops, soil exposures, vegetation, agriculture, state of existing structures, back-filled excavations, etc.

Subsurface drainage conditions will often be reflected in the natural vegetation; e.g. tubular grasses, willow trees, certain types of poplars and other hydrophilic plants may indicate a shallow water table, or a line of indigenous thorn trees may indicate a dyke in semi-arid areas. Local agricultural practices may also be useful indicators of soil conditions, e.g. ground which is not freely draining will usually be made over to pasture rather than to cultivation. Local farmers or other residents may be able to provide useful information about the history of the site. Such information should not be underestimated because it is qualitative and 'non-engineering': the engineer who takes pains to discover and map a zone of disturbed soil across his site will seem foolish when a neighbouring farmer later tells him that there was once a canal there, since filled with town rubbish.

All existing structures on and adjacent to the site should be examined for signs of distress. Cracks in buildings or distortions in telephone lines or road surfacings will, more often than not, have been caused by subsoil conditions. It is often possible to interpret the nature of soil movements from the pattern of cracking in a building, e.g. a 'corners-down' pattern will either be associated with dome-shaped distortions due to the heaving of active clays, or with the peripheral settlement of foundations on the inundation of a weak soil, whereas X-cracks invariably result from seismic shocks.

Records of mining beneath the site (undermining) and of the distribution of buried services should be sought from the appropriate government or local authorities. Records of previous construction on similar sites in an established city may also be available from local libraries and museums.

All relevant data gleaned from the enquiries and site inspections should be plotted with all available geological data on a site plan of appropriate scale.

Soil profiling

Using the procedures outlined in Chapter 6, soil profiles should now be recorded all over the construction site. Wherever possible this should be done by descending augered trial holes of large diameter; where this is impossible, owing to the presence of a shallow water table, soil profiling is done from samples recovered from boreholes (Table 6.6) and the properties of the soils are deduced from the results of laboratory and *in situ* tests; these indirect procedures are described later. The spacing of trial holes usually varies from 50 m to as little as 10 m or even closer, depending on the geological complexity of the site. In certain circumstances it may be useful to lay out holes on a grid, but it is more economical to locate holes in relation to the geometry of the local geology, e.g. to the dip and strike of sedimentary strata, and to the plan geometry of the structures to be built. Thus a plan of the layout of the holes should be prepared in advance, taking into account all variations in topography and known geology and all relevant features observed during the site inspections or revealed by enquiry. But provision must always be made for additional holes to be put down as the work proceeds. On comparing the descriptions of adjacent holes it may be found that the same horizon recurs in similar sequence at comparable depths over part of the site, but that it changes abruptly in depth or thickness in other parts, or that new and significant horizons appear. Probably the first uniform part of the site was sufficiently described from the holes shown on the original layout plan. If the changes are important, additional holes will be needed to locate their limits. Where the contacts of vertically oriented discontinuities must be located, the horizontal distances between trial holes on either side of them are successively halved, and extra holes or probes of some sort (e.g. SPTs) are put down at these points. The number of holes or probes may sometimes be reduced by using geophysical methods to guide their location (Appendix C). The need to make deep bores or probes may also be reduced by good geological interpretations: if the geological succession can be established from observations at shallow depths, and if there are grounds for believing that it does not change over short distances, it may be possible to infer what lies between 20 and 50 m once it is known that the horizons between 10 and 15 m represent the top of the succession (Brink 1979). In those parts of the world not subjected to Pleistocene glaciation, the depth to identifiable geological formations seldom exceeds 10 m except in infilled valleys. In the highly variable soils of glaciated areas, however, the spacing of trial holes may have to be much closer and their depths much greater.

There are no hard and fast rules as to how many soil profiles will be sufficient, but a sufficient number of soil profiles must be recorded to give a three-dimensional picture of the subsoil throughout the site and to the depth of stress influence of the proposed structure: this will be entirely dictated by the local geology and by the requirements of the structure.

Sampling for indicator testing

Atterberg tests and grading analyses are so simple and inexpensive to perform that they always form part of the preliminary investigation. Disturbed samples—each weighing about 1 kg—are collected in bags from every horizon of widespread occurrence and from any horizons of particular interest or concern. If soil conditions are complex, as many as 10 may be collected from a trial hole 20 m deep. The test data not only help to confirm the correlations of horizons from one trial hole to another, but also serve to determine the soil class of each horizon in the Unified Classification from which a variety of empirical deductions may be made (Table 2.2, pp. 33–40). These deductions provide conservative values for design purposes.

EXAMPLE: Site investigation for the London Central YMCA (Willbourne 1972; Burland 1979)

The site investigation and foundation design for this £4.5 million building complex in the heart of London illustrates the importance of accurate soil profiling by means of visual and tactile observation in a trial hole of large diameter—a technique which is rapidly gaining favour even in areas of shallow water table such as London.

The structure includes a basement 16 m deep, a three-storey podium above ground level and, on top of this, four diamond-shaped hotel towers varying in height from 9 to 14 storeys. The basement is 75 m long by 50 m wide and covers the entire site; it has diaphragm walls around the perimeter which, during excavation, were supported by ground-anchors inserted at an angle of 40° to the horizontal and extending 12 m into the London Clay. The sports and leisure facilities housed in the basement require large open areas, and to achieve this each tower had to be founded on a maximum of eight columns grouped in pairs carrying up to 2200 tonnes each; they are supported on under-reamed piles.

Owing to the congested nature of the site before demolition of the pre-existing buildings, and the proliferation of their heavy foundations, the preliminary investigation had to be restricted initially to the drilling of one borehole of 150 mm (6 inch) diameter. This revealed transported soils 7 m thick overlying London Clay to a depth of about 25 m, and followed by the Woolwich and Reading beds. It was out of the question to found the piles in the London Clay as the magnitude of the loads would have necessitated enormous under-reams which would have overlapped with the adjacent piles. It was hoped that the underlying Woolwich and Reading beds might provide an adequate founding stratum, but very little was known about the geotechnical characteristics of these beds.

Once demolition had been completed on part of the site it was possible to extend the investigation and it was decided, with advice from the Building Research Station, that more useful information would be gained from sinking one 1 m diameter trial hole to a depth of more than 30 m than from a number of smaller diameter bores which could not be descended for inspection. During the course of recording the soil profile in this trial hole (Fig. 7.1) it was observed that the condition of the London Clay had deteriorated

FIG. 7.1. Visual inspection in a 1 m diameter trial hole revealed this soil profile beneath the YMCA site in Tottenham Court Road, London. (Photograph by Cement and Concrete Association, London.)

— SURFACE

VARIABLE FILL UNDERLAIN BY PEBBLE MARKER OF ALLUVIAL GRAVEL
(TRIAL HOLE CASED INTO LONDON CLAY).

6.9 m

MOIST, DARK GREY, IN PROFILE DARK GREY, FIRM TO STIFF, FISSURED,
SILTY CLAY; LONDON CLAY.

19.2 m

—100 mm LAYER OF MOIST, LIGHT YELLOW, LOOSE, INTACT, SILTY SAND.

—LONDON CLAY AS ABOVE.

21.4 m

—70 mm LAYER OF WET, DARK GREY, SOFT, INTACT, SANDY SILT.

22.8 m — { WET, DARK GREY, IN PROFILE DARK GREY, FIRM TO STIFF, FISSURED, SILTY CLAY
{ BECOMING MORE SILTY WITH DEPTH; LONDON CLAY
—100 mm LAYER OF WET, DARK GREY, LOOSE, INTACT, SILTY SAND.

MOIST, BROWN, IN PROFILE BROWN, STIFF, INTACT, SILTY CLAY;
LONDON CLAY.

25.7 m

{200 mm LAYER OF SLIGHTLY MOIST, RED, IN PROFILE RED MOTTLED BROWN AND
{YELLOW, VERY STIFF, INTACT, SILTY CLAY; WOOLWICH AND READING BEDS.

SLIGHTLY MOIST, ORANGE, IN PROFILE REDDISH BROWN BANDED AND MOTTLED
YELLOWISH BROWN, STIFF TO VERY STIFF, INTACT, SILTY CLAY;
WOOLWICH AND READING BEDS.

28.5 m

—50 mm LAYER OF DRY, LIGHT GREY, FIRM, INTACT, SILT.

29.5 m — {MOIST, BROWN, IN PROFILE LIGHT BROWN, STIFF TO VERY STIFF, INTACT,
{SILTY CLAY; WOOLWICH AND READING BEDS.

{MOIST, GREY, IN PROFILE GREY, STIFF, SLICKENSIDED, CLAY;
30.6 m {WOOLWICH AND READING BEDS.

{MOIST, RED, IN PROFILE RED, VERY STIFF, SLICKENSIDED, SILTY CLAY;
31.5 m {WOOLWICH AND READING BEDS.

NOTE: WATER TABLE PERCHED BETWEEN 21.4 m AND 22.9 m.

during a period of eight hours; the silty layers were water-bearing and the fissured clays softened very rapidly. The underlying mottled clays of the Woolwich and Reading beds, however, were very stiff in consistency and, contrary to the widespread belief that they would contain water under artesian pressure, they yielded no water: they would thus clearly accept large diameter under-reams which the London Clay would not. From the assessment of consistency it was estimated that a bearing pressure of about 1000 kPa could be used on these beds. This figure, which was used for design purposes, was subsequently confirmed with plate-loading tests at the bottom of the first pile holes to be augered: the clay at founding level exhibited a load-bearing capacity of > 3660 kPa with a very small settlement, thus giving a safe bearing pressure of 1000 kPa with a factor of safety of 3 against failure. Even with an allowable bearing pressure on the founding material as high as 1000 kPa, it was still necessary to rake the piles at 1 in 25 to avoid the overlapping of their pressure bulbs. As the lowermost basement slab was expected to rise with time due to relief of stress from the excavated soil and subsequent absorption of water by the clay, construction in the bottom storey had to be as flexible as possible to allow deformation to take place without distress to the other building components.

The fact that the decision to found on the Woolwich and Reading beds was taken on the very day that the soil profile was recorded emphasizes the value of *in situ* observation within a large-diameter trial hole.

DETAILED INVESTIGATION

In most cases the preliminary investigation procedures described above will provide sufficient data for design. Burland, Broms, and de Mello (1977) state that 'it is probably not overstating the case to say that in 95 cases out of 100, decisions as to the type and depth of foundations can be made directly on the basis of a knowledge of the soil profile and ground water conditions, set in the context of the local geology and tied in with local experience, plus a detailed and systematic description of the soil in each stratum in terms of its visual and tactile properties.' Sometimes detailed exploration must follow. But it would probably not be overstating the case to add that 95 per cent of site investigations need not go beyond this preliminary stage. Accordingly this chapter has concentrated on this stage. Detailed exploration should never be undertaken unless it is clearly necessary and economically justifiable. 'The strong appeal is made—do not rush into such [detailed] procedures or be blinded by the apparent exactitude of involved tests or mathematical interpretations. These do have great uses provided that they are used only when absolutely necessary and are applied with proper understanding' (Jennings and Brink 1978).

Detailed investigation is necessary only when:

(i) it has not been possible to describe the soil profile properly, and particularly when it has not been possible to assess the consistency of individual soil horizons;

(ii) the consistency of the soil at an economical depth of founding is critical in relation to the minimum required bearing pressures;

(iii) large dams are to be designed: these always require detailed investigation, and particularly the acquisition of data on permeability;

(iv) special soil problems requiring further investigation have been identified during the preliminary investigation (Chapter 5).

The preliminary investigation will have provided a clear definition of the constraints imposed by the site, and the engineer will now plan the detailed investigation programme specifically to solve these problems. Subsequent procedures will therefore vary greatly. A brief description of some of these follows; fuller details are given in Appendixes A and B.

If a detailed investigation has to be undertaken because of inadequate data on soil consistency, an attempt must be made to obtain good undisturbed samples for consolidometer, triaxial, or shear box tests in the laboratory (Appendix A). The sampling must be done with the utmost care; bad samples submitted to the laboratory as 'undisturbed' are worse than no samples at all. The best and cheapest method is to recover samples manually from the sides of trial holes.

Where it is impossible to obtain good undisturbed samples manually from a large diameter trial hole—as is all too often the case with sands and with very soft clays or silts from below the water table—resort must be made to an appropriate method of sampling from boreholes or to *in situ* testing (Appendix B).

In deep sandy soils below a shallow water table

Under these conditions it is not possible to sink an uncased borehole nor can undisturbed samples be recovered. The soil profile can be reconstructed, albeit inadequately, from wash-boring samples or from the spoils recovered from shell-and-auger holes. Consistency is best assessed from Dutch probe or pressuremeter tests carried out at appropriate depth intervals; where a continuous record is necessary, Dutch probe soundings may be recorded every 200 to 250 mm. Other more costly methods include plate-bearing or pile-loading tests. However, indicator samples are readily recovered and the standard penetration test (SPT) provides a reasonable indication of soil consistency. Where conditions are complex and consistency assessments are required at close intervals, the N-value can be recorded every 500 mm, but it is more useful to deepen the trial hole by wash-boring or shell-and-auger drilling and to measure the N-value at test points spaced at 1.5 m depth intervals.

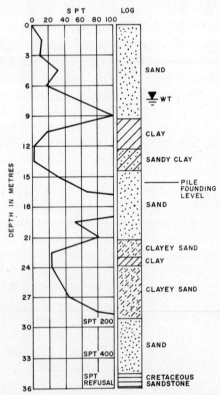

Fig. 7.2. Details of subsurface conditions below the Maharani Hotel, Durban, as revealed by standard penetration tests. (After Frankipile 1979.)

Investigations of this type were carried out for the Maharani Hotel in Durban (Frankipile 1979). The building is 32 storeys high and is situated on estuarine deposits overlying Cretaceous sandstone at a depth of 34 m. The result of SPT probes (Fig. 7.2) indicated that the foundations should be taken below the compressible clay and sandy clay layers present at depths 10 to 14 m. The solution adopted was to found the building on 122 driven tube piles of 560 mm diameter founded at a depth of 16.5 m to give a working load of 3400 kN per pile. In order to reduce the potentially damaging shocks of pile driving on the neighbouring structures, the piles were jetted through the dense sand layers above 9 m. Support was provided both by end-bearing in the dense sand below 16 m and by skin-friction along the pile shafts within the overlying materials.

In deep clayey or silty soils below a shallow water table

Casing would be required to support the sides of the borehole and a thin-walled Shelby tube, which is useful for sampling more cohesive materials, may not prove effective. Instead, a piston sampler may be needed to recover undisturbed samples of very soft clays or silts. Dutch probe or standard penetration tests do not yield very meaningful results in such soft materials, but Swedish vane or pressuremeter tests may provide information on soil consistency and strength at regular depth intervals. Similar results can be obtained at greater cost by plate-bearing or pile-loading tests. Settlement predictions can best be made from the results of consolidation tests carried out on undisturbed piston samples.

A typical investigation is described by Pike and Saurin (1952) for the site of a power station at Grangemouth in Scotland. Here silty estuarine deposits overlie stiff boulder clay or till at depths of 25 to 75 m. The soil profile was described from inexpensive wash-borings which were used to determine the depth to the till and to recover samples for indicator testing. Vane shear tests were carried out at regular intervals (Fig. 7.3). The foundation pressures for different parts of the complex varied considerably, and it became obvious at an early stage that no single type of foundation would suit the soil conditions. Three types of foundations were considered for the heavily loaded structures:

 (i) steel or concrete piles driven into the till;
 (ii) steel or concrete caissons sunk onto the till;
(iii) reinforced concrete cellular rafts.

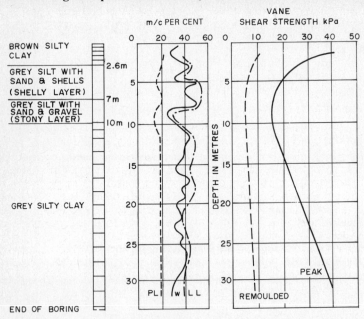

FIG. 7.3. Soil conditions at Grangemouth power station as revealed by wash-borings, indicator tests, and shear-vane tests. (After Pike and Saurin 1952; Skempton 1949.)

Preliminary estimates indicated that the cellular rafts would provide the cheapest solution. The reinforced concrete 'egg-crate' which was designed for this purpose was founded at a depth of 4.75 m, at which depth the mass of the excavated soil balanced the mass of the structure (23 100 tonnes). The cells of the concrete raft were kept dry by continuously pumping seepage water through well-points to ensure that no change in the pressure exerted on the soil beneath the foundations took place. This case history records a highly successful exploitation of 'earth buoyancy' for the support of a major structure.

For large dams

Site investigations for large dams always employ detailed procedures, and, particularly when the dam is to be built in an area underlain by deep unconsolidated deposits such as alluvium, these usually comprise four phases:

1. Once the site of the wall or embankment is selected by means of airphoto interpretation and field assessment (e.g. by seismic survey and exploratory borings) and the limitations imposed on the type of structure by the geometry of the valley and the availability of natural construction materials have been assessed by reconnaissance surveys, a relatively intensive primary drilling programme is planned. Rotary core drilling is usually dictated as the chief investigation method by the presence of a shallow water table in areas adjacent to the river floodplain and the need to penetrate into bedrock to a depth at least equal to the height of the dam. Drilling sites are selected so as to reveal expected fluctuations in the bedrock level, particular features of the geological structure such as faults, fracture zones, and intrusions which may have influenced the course of the river and the form of the valley, and also to show the distribution of different soils. It must be remembered that concrete dams which impose high foundation pressures or which transmit large arching stresses to the abutments usually have to be keyed into sound rock, but the outer granular shell of earth or rockfill dams can sometimes tolerate moderate settlements, and the design engineer may therefore elect to found these parts of the embankment above the bedrock level. The nature of the soils beneath the line of the dam also dictates the type of support required while excavating the foundations, and will often determine the type of impermeable cutoff installed below the embankment when this is founded above the bedrock surface. Triple-tube core-barrels (Fig. B.6, p. 286), are therefore used to ensure the best recoveries of unconsolidated material; these disturbed samples are usually supplemented by undisturbed tube samples. In alluvial sands which cannot be sampled satisfactorily, the *in situ* properties of the soil are assessed from the records of Dutch probes or standard penetration tests, between which the borehole is deepened by augering or wash-boring.

The primary boreholes are usually 50 to 100 m apart, and more than one line of holes may be selected if the wall or embankment is wide. The frequency of undisturbed samples depends on the complexity of subsoil conditions, but usually varies from one per metre to one per 5 metres. In extremely complex areas, successive tube samples are sometimes taken to provide a continuous record of the soil profile. Dutch probe tests usually involve continuous penetrations with readings taken every 200 mm, but with SPTs the N-value is usually recorded every 0.5 to 1.5 m.

The *in situ* permeability of all subsurface materials is regularly determined by well-pumping or falling-head tests in unconsolidated materials, or by water pressure tests in rock; in pressure tests, water-tight packers are used to isolate a particular section of the borehole. The stages tested in this way usually vary from 3 to 5 m in length.

2. The variations revealed by tests at the primary sites will determine the location of secondary sites to pinpoint geological contacts. In critical areas, secondary boreholes may be spaced as close as 10 m. If the early drilling results show that grout injection will be needed to prevent seepage beneath the dam, or to stabilize unconsolidated materials which may otherwise flow

or collapse into excavations, grout acceptance tests are usually undertaken. These are conducted in various ways, but an effective method is to drill three boreholes at the apexes and one in the centre of several equilateral triangles of different size; the central borehole is subjected to water pressure testing. The outer boreholes are then injected with various types of grout in different consistencies and at different pressures; this injection proceeds in stages until the full depth of the borehole has been grouted. Thereafter water pressure tests are again carried out in the central boreholes. In this way the degree of sealing achieved by the different types of grout and the different injection procedures can be related to distances between injection points, which facilitates the specification of a grouting programme during construction of the dam. Cement grout is normally used to seal fractured bedrock but chemical grouts or cement bentonite mixtures are more effective in sealing and stabilizing soils. The even injection of grout into soil is often difficult to control and is best achieved by the *tube-a-manchette* technique in which the grout is forced through a perforated casing surrounded by a membrane; the membrane prevents the soil from entering the casing and is easily ruptured once pressure is applied to inject the grout. Another method applicable in cohesionless sands is *disturbance grouting*, in which water is pumped under pressure from one borehole to another, usually at a specific level which is controlled by using a casing with a permeable segment; the flow of water raises sand into suspension, and cement grout is quickly injected into the mass before settling can occur.

3.   Where the shear strength and settlement characteristics of sub-foundation materials are critical to the long-term performance of the structure, or the possibility exists of a slip-circle failure through the base of an earth dam, additional *in situ* tests, such as plate-bearing or pressure-meter tests, are sometimes carried out in critical areas. These tests are particularly useful to augment the results of small-scale laboratory tests on undisturbed samples when the engineer must decide whether or not to remove clayey soils before founding the outer zones of an earth embankment. They also help the engineer to assess the effects of the differential strains produced by the small settlement of a deep cutoff and the much larger settlement of the adjacent embankment zones founded at shallow depth.

4.   Where the dam is to be of earthfill construction, bulk samples are taken of any soils to be removed from the cutoff and foundation excavations, which may be suitable for use in the embankment. These samples are tested in the laboratory in the same way as those recovered from the 'borrow areas' selected as the main sources of construction material for the various zones of the dam. Essential tests include triaxial tests on compacted material for shear strength calculations, moisture–density determinations to assess its compaction characteristics, and permeability tests to determine the water-retaining or drainage characteristics of the compacted material.

On the results of all these investigations the engineer makes several fundamental decisions affecting the final design of the dam. These require information to be made available *in good time* and include:

(i) The type of structure and cutoff best suited to local conditions—for example, where deep saturated sandy alluvium fills a wide valley and suitable natural construction materials are available, the engineer may decide to use an earth embankment founded at shallow depth with a narrow impermeable concrete cutoff, such as a diaphragm wall or sheet piles, taken down to bedrock. However, on some clayey soils the risk of major settlement or shear failure within an earth embankment might be so high as to necessitate a concrete gravity dam with deep foundations placed on bedrock.

(ii) The method of excavating the cutoff and foundation trenches—in saturated sand, open excavation might prove impossible without resort to expensive measures such as freezing the soils in the sides of the trenches; in this case a diaphragm wall might be more economic, or an impermeable clay blanket placed in the dam basin upstream of the embankment.

(iii) The design of an earth embankment—when the foundation excavations constitute a major source of construction material, the findings of the site investigation will control the zoning of the embankment and the safe slopes of its upper and lower faces.

INTERPRETATION OF SITE INVESTIGATION DATA

For the engineer to interpret the information assembled during a preliminary investigation he must be satisfied that the investigation has been carried out by an experienced soil profiler. In interpreting laboratory data from a detailed investigation he must be satisfied that the samples tested were sufficiently undisturbed. In interpreting the results of *in situ* testing techniques he must remember that these have measured things as they existed at the time of observation and that the results are ordinarily concerned only with the undrained condition, e.g. the undrained shear strength. In the design of a major structure, the interpretation of site investigation data is an art that relies heavily on engineering judgement. It relies heavily, too, on a close network of observation points, and as little as possible on interpolation between them.

Recommended reading

British Standards Institution (1981). Code of Practice for Site Investigations. BS 5930. London.

8. Medium scale surveys with the aid of remote sensing and geophysical techniques

In engineering projects that cover large areas, such as housing schemes and the provision of services for all aspects of urban development, it is usually not economic to collect detailed point data, of the type described in Chapter 7, from all over the project area. In his initial evaluation the engineer is not usually concerned with the choice of specific types of foundation treatment, since the precise nature of the structures and engineering works is not usually known at this stage; he is concerned rather with broad planning decisions on the zoning of different kinds of construction and hence with the distribution of various types of problem and their likely severity, and with the distribution of materials which can be used in the development (e.g. those suitable for road construction). In short, the shift in emphasis from *detailed design* to *general planning* justifies the use of cheaper and quicker procedures, a reduction in the density of site investigations by trial holes, test pits, boreholes, or penetration probes, and the use of techniques that infer the properties of the areas between widely spaced observation points from their similarities to the terrain at those points (Partridge, Brink, and Mallows 1973). Remote sensing and geophysical techniques are useful, sometimes essential, aids in this process, but much may be achieved through good field observation and common sense. The results are usually presented on a map. The reliability of the boundaries between land units will vary according to the method of their interpretation and the magnitude of the differences in their soil properties. Maps of this kind are useful for site selection and planning, and will often indicate the type and location of the detailed investigations required for individual structures.

The procedures discussed in this chapter are suitable for urban planning at local authority level, e.g.:

(i) planning the zoning of urban functions according to the suitability of soils for the associated structures (e.g. selecting which zones are suitable for high-rise structures, conventional housing or light and heavy industry etc.);

(ii) early identification of which types of foundation treatment will be required for new townships or housing schemes;

(iii) recognition of potential soil problems that will affect piped services, urban roads, or elevated motorways; and

(iv) route selection for shallow tunnels (e.g. for underground railways and sewers).

All the above applications require semi-detailed information to be presented at map scales from 1:1000 to 1:10 000. It is accepted that the information will be considerably less comprehensive and precise than that produced by the investigations discussed in Chapter 7, and that supplementary studies will usually precede the planning and design of specific structures.

SOME BASIC PROCEDURES

Semi-detailed maps are intended to distinguish areas of more or less uniform soil profile, or that contain groups of different profiles that display similar engineering properties, with moderate accuracy and for moderate cost. Mapping procedures fall into one of two groups:

The use of simple analogy

The survey area is broadly subdivided into land units that show some uniformity in their readily visible properties (e.g. surface landform, natural plant associations). There may be separate occurrences of each kind of unit in different parts of the area. Thereafter, each *kind* of subdivision is examined in detail at a few sites and it is assumed that the information from these sites is applicable to the whole of the unit they represent. Mapping units of this kind may show considerable internal variation. This procedure forms the basis of terrain classification (Chapter 10).

An example of this type of procedure is shown in Plate 4. The stereogram was prepared from panchromatic aerial photographs. The annotations attempted to subdivide the terrain into land units distinguished by their landform, micro-relief, and vegetation. When the annotation was complete, test pits were sunk at sample sites in those units considered likely to have a significant soil cover. They indicated that there was a consistent sequence of different soil horizons in each unit, but that the horizons showed considerable variations in texture and thickness.

Network sampling and interpolation

The area is sampled at a network of sites, either located on a regular grid or at positions chosen by the surveyor to sample the local range of conditions as fully as possible. The extent of each of the soil units thus revealed is then determined by matching sampled to unsampled areas, according to their tone patterns on aerial photographs, or their spectral signatures on remote sensing imagery, or on their physical properties (see Appendix C on remote sensing techniques and Appendix D on geophysical techniques). Boundaries are then interpolated between dissimilar units.

Plate 5 illustrates this second approach, with its considerably denser network of test sites. The distinction between the windblown sand (WS) on the upper slopes of the right bank of the river and the more clayey hillwash (H$_G$) on the left bank would not have been clear from a preliminary airphoto interpretation, but once they had been identified by detailed field observations the limits of their occurrence could be mapped quite readily. Likewise, the two different types of alluvium on the river floodplain could be distinguished only after detailed field probing. The presence of a weathered dyke (arrowed on the left bank) had been suspected from the occasional presence of diabase boulders on the lower slope, but its position was confirmed only by combining careful airphoto interpetation with magnetometer traverses.

Except in relatively simple terrain, the first (analogue) approach will often produce mapping units that contain too many soil differences to provide a basis for meaningful planning at the map scales usual for urban development projects. The analogue approach is likely to be more useful in regional planning, at mapping scales usually smaller than 1:50 000 (Chapter 10). The remainder of this chapter is concerned with the second alternative.

REQUIREMENTS OF THE FIELD INVESTIGATION

Preliminary studies

At the outset the surveyor should consult all existing geological and soil maps and other publications on the area, and attempt to unearth the findings of any other local investigations even if their purpose was only of marginal relevance to the project in hand. This may prove to be onerous, but it may eliminate expensive duplication. Specific site studies may already be available for urban areas which have already undergone some development. Chapter 12 discusses the co-ordination and storage of such information.

This sequence of commonsense preliminaries culminates in a comprehensive field reconnaissance which traverses all available roads and tracks to inspect all accessible exposures. This initial reconnaissance provides a general acquaintance with the area; even more importantly, fresh impressions of a particular tract of terrain often show the associative criteria (e.g. between soils or rock types and landforms, land-use, or vegetation types) in terms of which soil boundaries may be interpolated later. The soil surveyor should have equipped himself with aerial photographs, and have familiarized himself with their layout and the broad patterns they display. Various types of imagery are discussed in Appendix C, and the types most suitable for medium scale surveys are considered later in this chapter. He will probably attempt a preliminary map after his preliminary reconnaissance.

Soil profiling and sampling

Network sampling and interpolation require higher sampling densities, i.e. soil profiles recorded in trial holes, test pits, and boreholes, than are usual in the analogue approach, and there may be problems in acquiring sufficient data. Observation points may be located either on a grid, or by means of stratified random sampling (Smith and Atkinson 1975) which is preferred in complex areas in which the boundaries of mapping units are likely to display very irregular shapes. An average spacing of 100 to 200 m may suffice under normal circumstances, but, where rapid changes occur (e.g. across a weathered dyke or small drainage features), may have to be reduced to as little as 10 m. Conversely, where conditions are very uniform over large areas, 1 to 5 test sites per square kilometre may be enough. A rough rule of thumb holds that the spacing of test sites should not exceed a third of the minimum dimension of a mapping unit; in any event, there should be at least three profile points in each mapped unit, and other points near its boundaries are useful to determine the nature of the transition zone and the degree of internal variation. The final decisions on sampling density must be based upon experience of earlier surveys, or of earlier stages in the current survey. Since getting into the field is often a large part of the field costs, it is better to have too many test sites than too few.

In urban planning for lightly loaded residential structures, soil profiling to 6 m is usually deep enough; active clays below this depth seldom produce significant surface movements. However, once it is clear that there will be no problems except that of differential settlement, on soils with a collapsible fabric or on compressible silts and clays (Chapter 5), investigations for light residential structures may be restricted to the uppermost 1 to 2 m below founding level. In contrast, where there is a possibility of sinkhole development in soluble rocks, probing should be carried down to bedrock or to the lowest level of the permanent water table. Where specific provision must be made for heavy structures, investigations should normally include the entire unconsolidated soil mantle down to bedrock (Chapter 7), but such detailed investigations will not normally be included in the initial soil survey for urban planning.

The number of soil samples for laboratory testing must also be decided. One could specify at least three samples from every kind of horizon in every mapping unit, but this would often be excessive, since many horizons occur in more than one mapping unit; it will be more economical to collect one sample per unit from every horizon that may cause engineering problems. The anticipated engineering problem, or the potential use of the material, determines the type and size of the sample (Appendix A). Ideally, every sample of undisturbed soil should be accompanied by a disturbed sample of the same soil, so that the results of rigorous soil mechanics tests, which, because of their cost, can be carried out only on key samples recovered at infrequent intervals, may be correlated with the basic index properties of the different soils encountered (Chapter 2).

Once the laboratory results on these samples are available, the adequacy of the field programme must be re-assessed. If the correlation between field observation and laboratory data is unsatisfactory, or if materials of one or more key horizons show a high degree of variability (e.g. a coefficient of variation of more than 25 per cent, or exceeding that pre-selected for the survey), supplementary field work and testing should be undertaken to localize the relevant changes and inconsistencies. Where the coefficient of variation is less than 25 per cent, the values of the relevant samples may usually be taken to represent all other occurrences of the soil horizon.

A new mapping unit is identified where the sequence of horizons changes or where there is a significant change in the physical properties of one or more horizons (e.g. in texture, consistency, or structure). Provided that soil profiles have been recorded systematically (Chapter 6), samples from horizons defined in similar terms usually display similar mechanical characteristics; ideally any soil unit defined on a particular sequence of horizons may be expected to show only a limited range of engineering properties. In practice there are usually small lateral variations, both in horizon thickness and in horizon properties. If their cumulative effect results in significant variations within a unit then it may need further subdivision.

Initially, surveys look for the boundaries between mapping units by shallow borings with a hand-auger along traverses between test sites. In areas where superficial changes are gradational, or where the cost of closely spaced traverses is not acceptable, the interpreter will use air photographs or remote sensing imagery.† All the field data are carefully compiled onto the images, and the surveyor analyses them stereoscopically; he looks for features common to all the sites that lie on the same unit, and for dissimilarities between the features of sites on different units. It will often be possible to amend the initial interpretation on the evidence of the field data and to subdivide the mapping units by looking for more subtle variations in the images to match the ground observations (Plate 5). The refinement of boundaries may be aided by techniques of image enhancement (Appendix C). Whenever possible, boundaries interpolated in this way should be located and confirmed in the field.

Appendix C describes the relative advantages of different types of imagery and the basic techniques for their interpretation. At the map scales of less than 1:10 000 which are required for the intermediate levels of detail required here, it is unlikely that remote sensing aids other than conventional aerial photographs (either panchromatic or colour), infrared photography, multi-spectral photography, or infrared linescan imagery will prove of assistance in interpolating boundaries between soil profile types. Indeed, unless special missions have been flown, the only imagery available

---

† The Webster–Beckett rule states that 'if the position of a soil boundary cannot be shown to be related to any observable feature of land use, land cover, or land form, then its exact location is probably not crucial either'; this is often, though not always, true.

will probably be conventional aerial photographs, often at scales very much smaller than those appropriate to detailed applications. If this is the case, and the photographic negatives are of high quality, the interpreter should endeavour to obtain enlargements of the photographs as close as possible to the scale of final map presentation. Where doubts still exist regarding the position of critical boundaries, it may be possible to interpolate them by geophysical techniques (Appendix D).

The special field investigations undertaken in areas in which sinkholes and subsidence may be anticipated are described in Chapter 5. When mapping such areas at intermediate scale, the exploratory drilling upon which assessments of stability are based should not be undertaken until the completion of the fieldwork described above and the completion of preliminary geophysical surveys. In this way the distribution of deep soil cover, outcrop areas, gravimetric anomalies, and the level of the water table (Fig. 8.1 and Appendix D) can be taken into account in selecting the drilling positions which would best sample the range of conditions present in a particular area, and best calibrate the geophysical observations. If possible, drilling should be carried out on all positive gravimetric anomalies, and water levels should be recorded in each separate groundwater compartment. Solid rock should be penetrated for at least 7 m in each borehole. Rates of drill penetration should be recorded and accurate borehole logs should be prepared from careful examination of the cores or cuttings (Chapter 6 and Appendix E). Where groundwater compartments are bounded by structural features such as dykes or silicified faults, the positions of these should also, where possible, be confirmed by drilling.

PRESENTATION OF DATA

The field and laboratory programmes produce a set of point data. The plan positions of each profile and the vertical succession of soil horizons and sampling points in each profile are known, so effectively the data are recorded in three dimensions.

The first step is to decide how to classify the information in terms that are relevant to the project. It is usually convenient to define mapping units so that each is occupied by a particular succession of soil horizons, each with a particular range of engineering properties. Where possible, the succession of horizons should include bedrock, or its weathering products, and so the mapping units often comprise vertical sequences of three or more significant soil materials.

These data can be presented in a variety of ways. In one form of presentation each mapping unit is identified by means of a number or letter. The material and origin of each horizon can then be listed in the legend, together with a summary of its more important properties (Fig. 9.1, p. 198). This technique is particularly useful when the map must show other data (e.g. contours and cadastral information) as well. The field observation points should be marked and it is sometimes useful to reproduce small simplified soil profile columns next to the relevant points. There may be problems in transfering data from airphotos or other images to the map, because of differences in scale and image distortions, but these can be adjusted with instruments such as the Stereosketch or Zoom Transferoscope which match the relevant images to facilitate direct transfer.

Another technique is to superimpose appropriate symbols or hatchings for each horizon; bedrock geology or parent material is often shown in colour. Each mapping unit is labelled by a number or a letter as suggested above, or by a sequence of letters to summarize significant properties such as texture or origin (e.g. H_s/RS may denote hillwash or fine colluvium derived from sandstone overlying residual sandstone; SSi/SC may indicate silty sand overlying clayey sand). However, when many soil horizons are superimposed a confusing juxtaposition of symbols may result.

Alternatively, horizon-sequence units may be grouped on specific engineering properties. These may be shown by means of separate subscripts or different symbols, but it is often less confusing to present them on a separate map. Such a map might present data on excavation characteristics, seepage areas, zones potentially subject to sinkhole development, depth to water table, potentially unstable slopes, and sources of natural construction materials, without confusing the basic information on soil distribution (Fig. 8.2).

An interpretive map may present specific ranges of engineering properties in a classification designed for planning purposes. This usually requires the regrouping of soil profiles in terms of their engineering properties and often involves some manipulation of data: old boundaries are redrawn and new ones added. The range

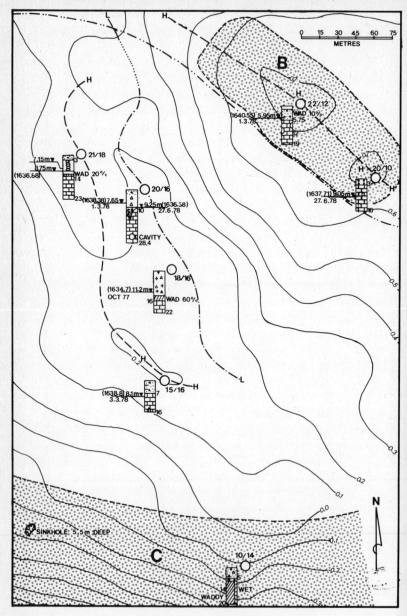

FIG. 8.1. Potentially unstable zones in an area underlain by cavernous soluble rocks can be delineated by a gravimetric survey; note the small sinkhole near the bottom left.

PLATE 1. Land facet annotations over a key area representative of a land system.

PLATE 2. The two land systems on which facet annotations have been drawn occur on the same bedrock but show markedly different relief and soils; they are separated by a bold line.

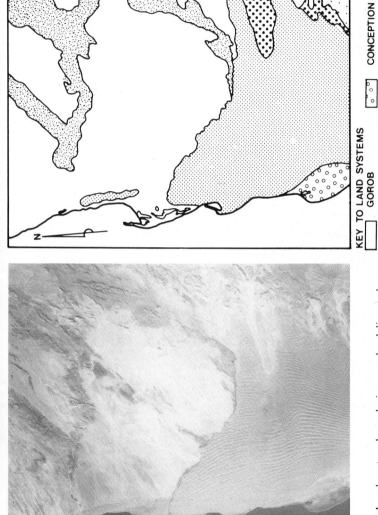

KEY TO LAND SYSTEMS

GOROB CONCEPTION

KHAN ABBABIS

SANDWICH TSONDABVLEI

PLATE 3. Land system boundaries may be delineated on small-scale photographs or other imagery from satellites.

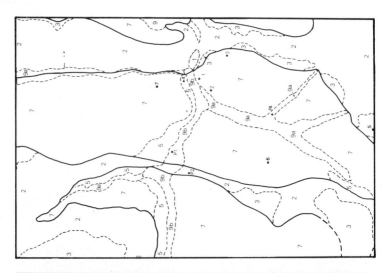

1: Solid outcrop of dolerite

2: Dolerite boulders in matrix of residual dolerite

3: Pediment deposit/residual dolerite

4: Slumped doleritic debris/sedimentary rock

5: Steep outcrop of sedimentary rock sporadically mantled by pediment deposit

6: Slumped sedimentary debris/sedimentary rock

7: Pediment deposit/sedimentary rock

8: Pediment deposit/dolerite dyke

9a: Clayey alluvium

9b: Sandy alluvium

9c: Bouldery alluvium

⁓⁓⁓ Geological contact

- - - Inferred geological contact

⁓⁓⁓ Land facet boundary

f — — —f Fault

•4 Test pit (soil profile recorded)

•P5 Exposure (soil profile recorded)

PLATE 4. An area may be subdivided into simple soil units by airphoto interpretation with field checking at a few points—these units are of comparable size to the land facets of Chapter 10. (By permission of O'Connell, Manthé, and Partners, Inc. and the Transkei Department of Agriculture and Forestry: Transkei Rural Water Supply Scheme.)

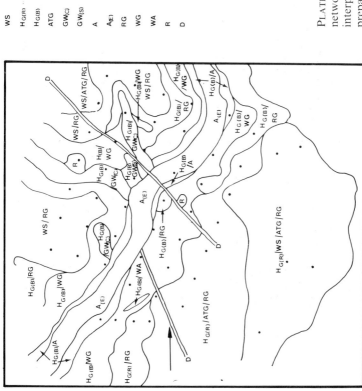

WS	WINDBLOWN SAND
$H_{G(R)}$	RED HILLWASH EX GRANITE (SL.CLAYEY SAND)
$H_{G(B)}$	BROWN HILLWASH EX GRANITE (SILTY SAND)
ATG	ALLUVIAL TERRACE GRAVEL
$GW_{(C)}$	GULLYWASH (CLAYEY – EXPANSIVE)
$GW_{(S)}$	GULLYWASH (SANDY)
A	ALLUVIUM (CLAYEY – NON-EXPANSIVE)
$A_{(E)}$	ALLUVIUM (CLAYEY- EXPANSIVE)
RG	RESIDUAL GRANITE (GRAVELLY)
WG	WEATHERED GRANITE
WA	WEATHERED AMPHIBOLITE
R	GRANITE OUTCROP
D	DIABASE DYKE

PLATE 5. Interpolation of soil boundaries within a network of sampling points by airphoto interpretation provides an alternative method of preparing a soil map for more detailed planning.

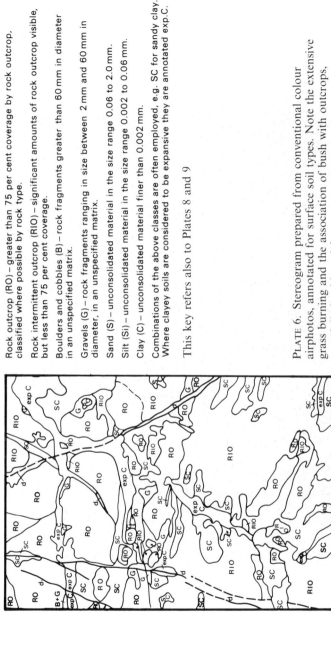

Rock outcrop (RO) – greater than 75 per cent coverage by rock outcrop, classified where possible by rock type.

Rock intermittent outcrop (RIO) – significant amounts of rock outcrop visible, but less than 75 per cent coverage.

Boulders and cobbles (B) – rock fragments greater than 60 mm in diameter in an unspecified matrix.

Gravels (G) – rock fragments ranging in size between 2 mm and 60 mm in diameter, in an unspecified matrix.

Sand (S) – unconsolidated material in the size range 0.06 to 2.0 mm.

Silt (Si) – unconsolidated material in the size range 0.002 to 0.06 mm.

Clay (C) – unconsolidated material finer than 0.002 mm.

Combinations of the above classes are often employed, e.g. SC for sandy clay. Where clayey soils are considered to be expansive they are annotated exp.C.

This key refers also to Plates 8 and 9

PLATE 6. Stereogram prepared from conventional colour airphotos, annotated for surface soil types. Note the extensive grass burning and the association of bush with outcrops, especially bouldery dykes.

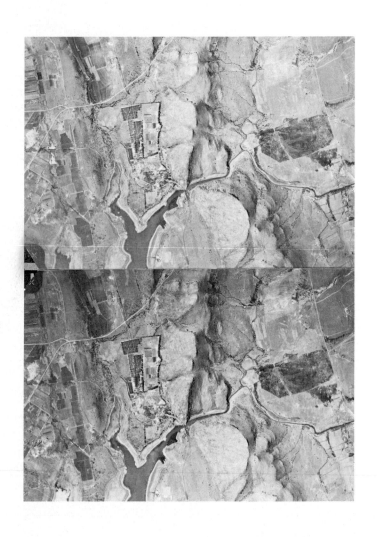

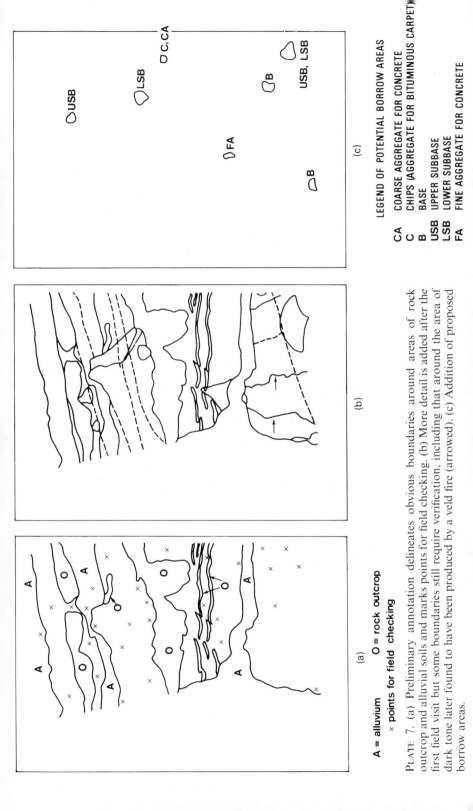

(a)

A = alluvium   O = rock outcrop
× points for field checking

(b)

(c)

LEGEND OF POTENTIAL BORROW AREAS

CA   COARSE AGGREGATE FOR CONCRETE
C   CHIPS (AGGREGATE FOR BITUMINOUS CARPET)
B   BASE
USB   UPPER SUBBASE
LSB   LOWER SUBBASE
FA   FINE AGGREGATE FOR CONCRETE

PLATE 7. (a) Preliminary annotation delineates obvious boundaries around areas of rock outcrop and alluvial soils and marks points for field checking. (b) More detail is added after the first field visit but some boundaries still require verification, including that around the area of dark tone later found to have been produced by a veld fire (arrowed). (c) Addition of proposed borrow areas.

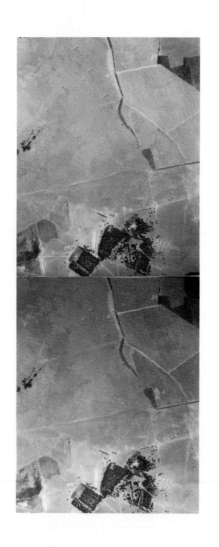

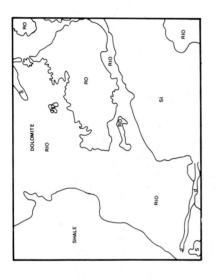

PLATE 8. Stereogram prepared from false colour airphotos, annotated for surface soil types. Note the red spectral signature of healthy trees and the clear definition of discontinuous rock outcrops.

For key, see Plate 6

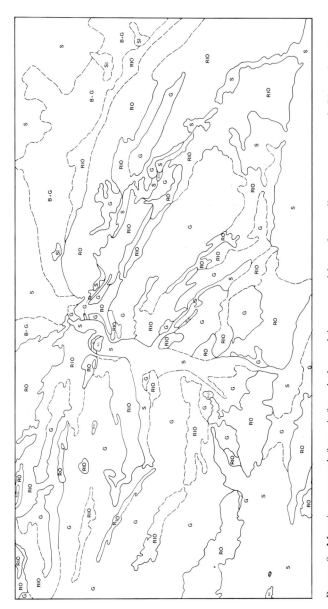

PLATE 9. Mosaic prepared from infrared thermal imagery with surface soil types annotated. Note the clear definition of the strike of shallow sub-outcrops and differentiation of well-drained granular soils from poorly drained alluvium.
For key, see Plate 6

PLATE 10. Portion of a multi-spectral colour-composite image prepared from digital data recorded by the Landsat II satellite. The accompanying annotations show land system boundaries. Note the very clear distinction between the clay residual from mafic rocks of the OP land system and the adjoining sandy soils characterizing the granite terrain of the HK land system

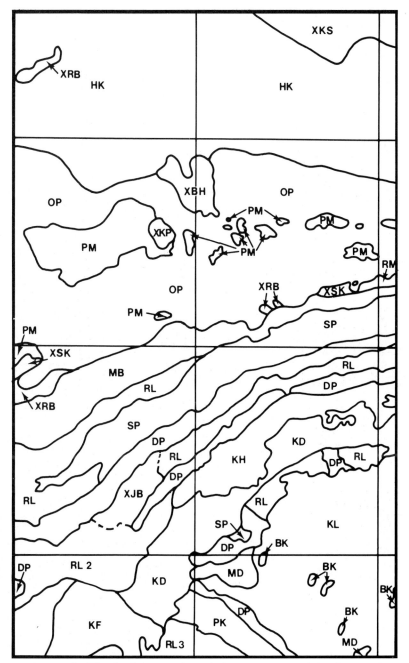

to the north. The land system boundaries were mapped by interpreting
panchromatic aerial photographs, supplemented by field checking—there is a
remarkable coincidence between these boundaries and the major changes in
spectral signature on the image. (Image prepared by Spectral Africa Ltd.)

PLATE 11. Stereogram constructed from standard panchromatic aerial photographs, annotated to show different soil profile units; the accompanying table summarizes the elements of the photographic image for each unit.

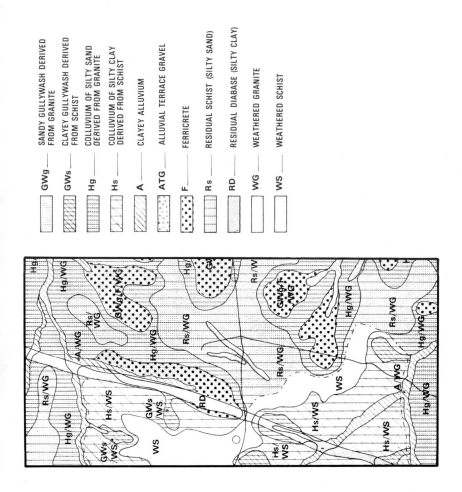

GWg	—	SANDY GULLYWASH DERIVED FROM GRANITE
GWs	—	CLAYEY GULLYWASH DERIVED FROM SCHIST
Hg	—	COLLUVIUM OF SILTY SAND DERIVED FROM GRANITE
Hs	—	COLLUVIUM OF SILTY CLAY DERIVED FROM SCHIST
A	—	CLAYEY ALLUVIUM
ATG	—	ALLUVIAL TERRACE GRAVEL
F	—	FERRICRETE
Rs	—	RESIDUAL SCHIST (SILTY SAND)
RD	—	RESIDUAL DIABASE (SILTY CLAY)
WG	—	WEATHERED GRANITE
WS	—	WEATHERED SCHIST

Soil origin	Tone	Texture	Shape and internal pattern		Stereoscopic appearance	Associative characteristic
			Shape and Structural Pattern	Drainage form +		
Rs/WG	Light grey	Moderately fine	Amorphous, showing occasional angular jointing pattern in igneous rocks, and foliation lines in schists	Absent except on old erosion surface where dislocated, with occasional pans	Gently sloping, convex	Some agriculture
Hg/WG	Medium grey with occasional dark patches	Medium, sometimes mottled	Amorphous around hill crests, no SP	Sub-parallel/radial, low density, good integration	Gently sloping, convex	Patches of low bush and crops
G	Light grey with heavy shadows	Medium	Circular with angular jointing pattern	—	Steep, with substantial relief	Quarries
G	Almost white	Moderately fine	Amorphous showing occasional angular jointing pattern	—	Gently sloping, convex	Absence of vegetation
GWg/F/WG	Medium grey, sometimes patchy	Fine, except where rilled at head	Hemi-lemniscate, no SP	Centripetal/sub-parallel at head. Dense but poorly integrated	Gently sloping, concave	Tall grass, crops, sand quarries
ATG/WG	Light grey	Medium	Linear/lenticular, no SP	—	Gently sloping, with steep sides	Trees
A/WG	Medium grey	Fine, except where gullied	Linear, sinuous, no SP	Sub-parallel when gullied; Moderately dense but poorly integrated	Gently sloping, except where incised by stream	Dams, tall grass
Hg/WG	Medium grey with occasional dark patches	Medium, sometimes mottled	Ring-shaped, no SP	—	Moderate slope, concave	Patches of low bush, sand quarries
P/WG	Medium grey	Very even, fine	Circular, no SP	—	Virtually flat	Standing water, beach line, tall grass
RD	Medium grey to black	Mottled, moderately coarse	Linear with occasional angular jointing pattern	—	Broken low ridge	Trees
Hq/WG	Medium grey	Medium	Broad belt below marginal escarpments, no SP	Sub-parallel, low density, good integration	Gently sloping, concave	Crops

+ Includes pattern, density and degree of integration

LEGEND

~0.3~	BOUGUER GRAVITY CONTOURS AT 0,1 mgal INTERVAL
—..—´	ASSUMED APPROXIMATE POSITION OF DIABASE SILL
H——H	GRAVITY RIDGE
L—.~.—L	GRAVITY TROUGH
(H)	GRAVITY HIGH
(L)	GRAVITY LOW
⬭	SINKHOLE
⬭	POTENTIALLY UNSTABLE AREA

1641,39 8,7m▼
27.6.78

WATER REST LEVEL BELOW GROUND LEVEL WITH HEIGHT
ABOVE SEA LEVEL IN PARENTHESIS; DATE OF MEASUREMENT.

◯21/10	BOREHOLE IDENTITY (ROW/COLUMN)

⬚	SANDY
⬚	CLAYEY
⬚	CHERT FRAGMENTS
⬚	HEAVILY FERRUGINIZED
⬚	LIGHTLY FERRUGINIZED
⬚	MUCH CHERT
⬚	HIGH WAD CONTENT (%)
⬚	VERY HIGH WAD CONTENT (> 20%)
⬚	DOLOMITE

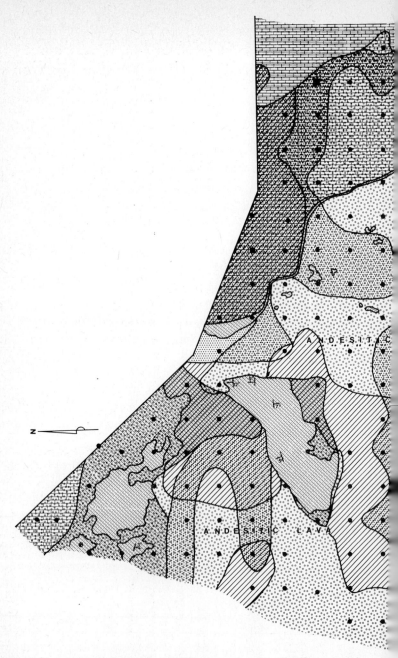

FIG. 8.2. A map showing the distribution of soils of different origins provides a useful first stage in identifying areas which may present problems in township development.

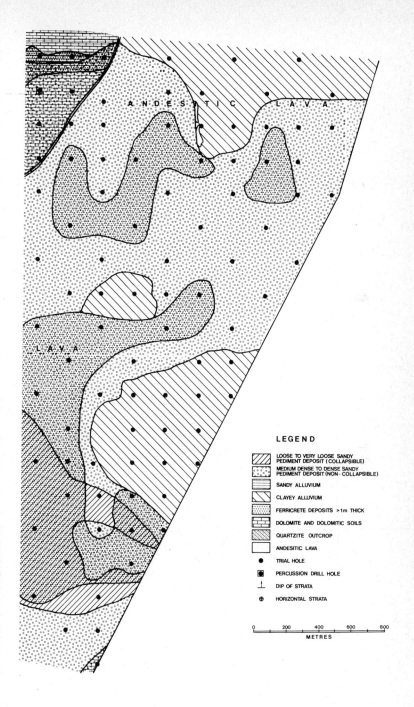

LEGEND

▨	LOOSE TO VERY LOOSE SANDY PEDIMENT DEPOSIT (COLLAPSIBLE)
⋯	MEDIUM DENSE TO DENSE SANDY PEDIMENT DEPOSIT (NON - COLLAPSIBLE)
▦	SANDY ALLUVIUM
▨	CLAYEY ALLUVIUM
▨	FERRICRETE DEPOSITS >1m THICK
▦	DOLOMITE AND DOLOMITIC SOILS
▧	QUARTZITE OUTCROP
☐	ANDESITIC LAVA
●	TRIAL HOLE
▣	PERCUSSION DRILL HOLE
⊥	DIP OF STRATA
⊕	HORIZONTAL STRATA

ANDESITIC LAVA

LAVA

0   200   400   600   800
METRES

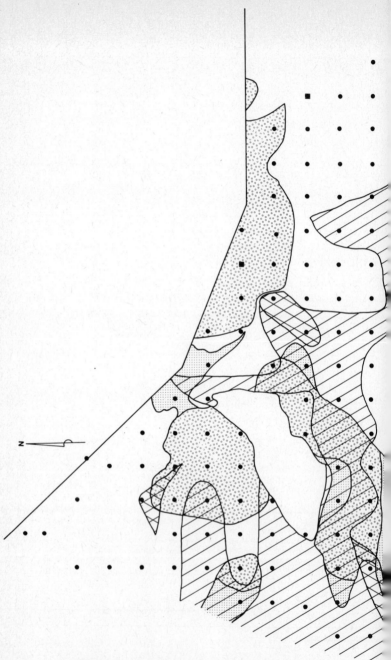

FIG. 8.3. A map showing the distribution of soils upon which houses may experience varying degrees of differential movement assists the planner in zoning the area for different urban uses.

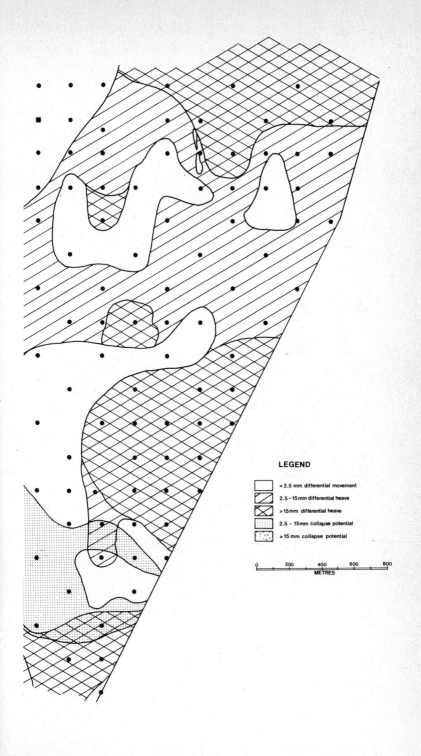

LEGEND

< 2.5 mm differential movement

2.5 – 15 mm differential heave

>15 mm differential heave

2.5 – 15 mm collapse potential

>15 mm collapse potential

0   200   400   600   800

METRES

of relevant properties for each unit will be calculated from the field and laboratory results, taking into account the cumulative effect of all soil horizons. For example, the potential expansiveness, depth and variation in thickness of successive clayey horizons may be used to calculate the range of heave which can be anticipated from the whole profile (Fig. 8.3) or the likely range of settlement under a particular foundation pressure can be generalized from the results of consolidometer testing.

A typical planning classification based on such characteristics indicates areas suitable for conventional housing, for medium and high-rise structures, for easy provision of underground services, and for roads and mass transit systems, as well as those areas which are best excluded from development and reserved for open space uses. Such interpretations would naturally depend on the type of development, the extent to which foundation treatment is considered practicable for different structures, and the degree to which other constraints restrict the alternatives. Interpretations of this type are best carried out in consultation with the planner.

The scale of the map will depend on the availability of base maps and the requirements of the project, as well as on the complexity of soil distributions. In highly complex areas, scales as large as 1:1000 may be required; more frequently scales range from 1:2500 to 1:5000, and information may be presented adequately at 1:10 000 when the mapping units are large. In complex terrain, carefully selected long-sections showing both vertical and lateral soil distributions may be helpful to the user.

THE USE OF OTHER DATA

Many areas, particularly near the urban concentrations of developed countries, are covered by official geological or pedological maps which are available from the appropriate government authorities. The results of private surveys are not usually available to outside parties, but, if recorded in a data bank, their existence is more easily traced and their examination can sometimes be negotiated (Chapter 12).

An understanding of the weathering and soil-forming processes which operate on different rocks under different climatic and geomorphological circumstances (Chapter 3) aids the interpretation of bedrock maps to predict the likely engineering properties of the soils in a particular area. Interpretation of such maps will also indicate where residual soils that may present problems occur beneath transported material. In the United Kingdom, geological maps also show different types of surface 'drift' (transported soil).

Information from agricultural soil maps can help in the preparation of engineering soil maps, although the depth of the records and the properties of interest to the agriculturist must invariably be augmented to elucidate engineering problems; the 1 to 2 m depth to which soils are examined in agricultural surveys may not include any residual soil or give an indication of its variation at depth. Agricultural soil maps nevertheless provide useful data on soil texture, mineralogy, and permeability in the near-surface horizons; the distributions evident in these materials are often a guide to variations at depth. In general, agricultural maps are likely to be prepared at scales more compatible with those used in medium-scale engineering soil maps than is the case with most geological maps.

DISCUSSION OF TECHNIQUES

In the approach to soil survey described in this chapter, the use of *associations* is of paramount importance. In all cases, changes are being identified on the basis of a group of attributes (e.g. surface form) which are, at best, only weakly associated with the properties which are of interest to the engineer. The soil surveyor hopes that the units he distinguishes in this way on their external appearance will reflect significant differences in the engineering properties of their associated soils. For example, much of the interpretation of soils from remote sensing imagery depends on the recognition of changes in surface terrain and natural plant associations. While these often reflect changes in geology and soils, this is not always the case, especially where significant variations are present deep below the surface. Similarly, most geophysical techniques measure certain conditions (such as the amount of interstitial water present in the soil) rather than the properties of the soil itself.

While an experienced interpreter can achieve a high degree of accuracy in interpolating soil boundaries by using these techniques in a familiar terrain, and where a close association exists between surface and sub-surface changes, he may achieve large discrepancies

TABLE 8.1

The uniformity of indicator test data within mapping units

Soil material	Description	Facet	Variant	Statistic	Liquid limit	Plasticity index	Linear shrinkage
A (BL)	Black alluvium	7	—	$\bar{x}$	58.5	37.1	13.7
				s	14.70	13.67	10.31
				V	0.2513	0.3685	0.7526
A (GR)	Grey alluvium	7	—	$\bar{x}$	51.0	32.4	13.4
				s	13.69	11.35	3.494
				V	0.2684	0.3503	0.2607
ATG	Alluvial terrace gravel	6	—	$\bar{x}$	37.3	19.7	8.43
				s	2.55	2.121	3.106
				V	0.0684	0.1077	0.3684
GW_g	Gully-wash derived from granite	5	1	$\bar{x}$	10.7	2.4	0.83
				s	8.659	2.098	0.8198
				V	0.8093	0.8742	0.9877
GW_s	Gully-wash derived from schist	5	2	$\bar{x}$	60.5	38.8	16.88
				s	4.93	3.661	1.036
				V	0.0815	0.0944	0.0614

F	Ferricrete	5	1	$\bar{x}$	19.8	6.3	2.64
				s	8.149	4.197	2.109
				V	0.4116	0.6662	0.7989
H_g	Fine colluvium derived from granite	2	1	$\bar{x}$	17.3	4.9	2.04
				s	8.979	3.701	1.044
				V	0.5190	0.7553	0.5118
H_s	Fine colluvium derived from schist	2	2	$\bar{x}$	40.9	22.3	10.93
				s	7.015	4.001	2.304
				V	0.1715	0.1794	0.2108
RS	Biogenic soil	1	1 and 3	$\bar{x}$	31.8	16.5	7.64
				s	6.997	5.785	2.488
				V	0.2200	0.3506	0.3257
RG	Residual granite	1	1	$\bar{x}$	44.1	22.6	9.78
				s	11.0	10.88	4.271
				V	0.2494	0.4814	0.4367
WG	Deeply weathered granite	1	3	$\bar{x}$	35.4	15.8	7.16
				s	3.47	4.517	2.147
				V	0.0980	0.2859	0.2999
WS	Residual schist	1	2	$\bar{x}$	44.9	16.4	7.44
				s	4.306	5.461	2.163
				V	0.0959	0.3330	0.2907

$\bar{x}$ = mean; s = standard deviation; V = coefficient of variation

under unfavourable circumstances. In these areas, the number of field sites to be examined and sampled must be increased in order to provide acceptable accuracy. Uniformity measures are sometimes applied to ascertain the degree of internal variation which may occur in mapping units delineated in this way. Table 8.1, which is based on a study specifically designed to assess the sorts of unit that an interpreter can map, indicates a fairly normal result and also illustrates the sort of trial or check that a mapper might make himself if he wished to know whether any of his units required further sampling to map out significant irregularities. The detailed sampling that produced Table 8.1 showed that a significant number of the units had coefficients of variation in excess of 25 per cent, taken as the useful upper limit for planning purposes; however, high degrees of variation may not matter if the total range is small, as is the case with the gully-wash derived from granite (GW_g) and the ferricrete (F).

EXAMPLE: soils investigation for an urban satellite area using network sampling and interpolation techniques

The area investigated covers about 1000 ha and is underlain by Precambrian andesitic lava, quartzite, and dolomite. Relief is low, and is associated chiefly with quartzite outcrops. The sampling network and distribution of the principal soil units and underlying lithologies are shown in Fig. 8.2.

Panchromatic aerial photographs at 1:10 000 scale proved effective in distinguishing the quartzite outcrops which separate the underlying lava from the overlying dolomite bedrock, and in recognizing the alluvial soils. The latter contain highly expansive clays and have high water tables during the wet season. The deep soils residual from lava are somewhat clayey and moderately expansive, but contain highly expansive zones where advanced decomposition has produced concentrations of smectite. The fine sandy colluvium derived from quartzite shows variations in consistency and void ratio, and includes zones with a collapsible fabric. Substantial ferricrete hardpans have formed in sideslope areas influenced by seasonally fluctuating water tables. The residual dolomite soils are often deep and could favour the development of sinkholes or subsidences, so percussion drilling in selected localities, to determine the depth to bedrock and to the local water table, was considered essential. In some areas, dolomite bedrock occurred above the phreatic surface.

The results were interpreted for lightly loaded or residential houses, and the provision of urban roads and piped services. Areas subject to potential settlement and heave were subdivided into zones as shown in Fig. 8.3.

In evaluating the engineering implications of these findings

(i) areas of rock outcrop, and units with a potential differential movement of less than 2.5 mm, were classified as suitable for housing development;

(ii) units with differential movement in the range 2.5 to 15 mm were proposed for conventional housing development, subject to the application of appropriate, relatively inexpensive, foundation and structural treatment (e.g. split construction, reinforcement, and strict moisture control in the vicinity of the foundations);

(iii) where severe differential movements are predicted, the costs of foundations may preclude housing, so that areas were zoned for heavier structures such as apartment blocks, shops, offices, and schools, since the expense of radical foundation treatment is more readily accommodated in such costly structures; alternatively, such areas could be reserved for open space uses;

(iv) restrictions were placed on building in the dolomitic areas in view of the local configuration of the bedrock and water table levels.

Difficulties were envisaged in excavating trenches for piped services in areas of rock outcrop or shallow hardpan ferricrete; however, the latter material is suitable for use in the controlled layers of roads. In areas where significant differential movements are likely, pipes require flexible joints, and roads need special foundation preparations, drainage, and cover design.

This investigation helped the urban planners entrusted with the layout of the area to balance the natural constraints imposed by soil conditions against other requirements.

9. The preparation of soil engineering maps for transportation routes using remote sensing techniques

In the planning of roads, railways, pipelines, and canals it may be necessary to identify specific problem areas, and to locate sources of natural construction materials, over long distances.

Traditional soil survey procedures involve sampling and testing at regular intervals along the proposed centreline and prospecting for construction materials by random pitting, based at best on the subjective hunches of technicians.

If a soil engineering map is available from the start, the tasks of centreline soil survey and materials prospecting can be planned systematically. When supplemented with data on relief, slope, and surface drainage, a soil engineering map can be used by the engineer in all stages of a project from initial planning to final construction and even in the monitoring of post-construction performance.

Soil maps for the design and construction of transportation routes will normally be at scales of 1:10 000 to 1:50 000. Under special circumstances, scales as small as 1:100 000 may be practicable, and the survey procedure then merges into that for regional soil survey (Chapter 11). Surveys of the large areas involved would be excessively costly by the field survey techniques appropriate to potential urban areas (Chapter 8), and greater use is made of remote sensing imagery and the analogue approach. Boundaries are drawn around areas of similar remote sensing pattern; the uniformity of such areas is then checked at a few points in a few of the areas which have been recognized as similar.

In humid areas, where the boundaries between mapping units are obscured by dense vegetation or by intense cultivation, aerial photographs are less useful and more intensive fieldwork must be undertaken. The use of side-looking radar imagery may assist in reducing the masking effects of dense vegetation. In general, however, it will be economic to concentrate expensive fieldwork in areas which remote sensing interpretation indicates as zones of high risk or high construction cost, for example for deep excavation or tunnels, or structures such as bridges, aqueducts, or high fills, or in areas where special soil problems or good construction materials are likely to occur.

DETAILS OF PROCEDURES

This chapter describes procedures for preparing soil engineering maps along the routes of roads or railways. The construction materials needed for roads and railways have been considered in Chapter 4, and particular soil problems in Chapter 5.

The approximate line of the route will have been fixed as a wide corridor, at the commencement of the project, by local political and economic considerations, or by the simple geometry of the straightest line that crosses no unbridgeable rivers or impassable hills. Regional soil surveys (Chapter 11) to reveal the main geotechnical constraints may have reduced the area under consideration to a strip within the corridor, perhaps 5 km on either side of the route. This is now to be mapped. Its limits will be decided in consultation with the design engineers. All available information about the geology, soils, topography, vegetation, climate, and so on should be assembled at the outset, with two sets of aerial photographs covering the strip. Photographic scales of 1:50 000 or smaller may prove suitable in arid areas, but it is not usually possible to delineate mapping units at scales much smaller than 1:30 000 in humid, densely vegetated areas. A mosaic is constructed from the central portions of each print from one set of aerial photographs. The centreline of the proposed route is inscribed on the mosaic, with chainage points at regular intervals. If a topographic map to a suitable scale (1:10 000 to 1:50 000) is not available as a base for the final soil engineering map, the mosaic may have to be used for this purpose too. The second set of prints should be numbered in sequence along each strip in the same corner of each print: A1, A2; B1, B2 etc.

The mosaic is then examined in conjunction with any available soil maps. The interpreter starts by attempting to delineate any discontinuities or changes in airphoto pattern that may correspond to soil or geological boundaries, e.g. prominent rock outcrops, alluvial floodplains, swamps, enclosed depressions, etc. (Plate 7(a)). He will not be sure what these tentative boundaries have separated until he makes a ground check, but even at this stage his local experience may help him to ignore such things as boundaries between burnt and unburnt veld, cultivated and virgin areas, and to concentrate on boundaries that may be significant. It is often useful to draw in the drainage pattern at this stage. Points which seem to represent significantly different soil conditions and which are easily accessible for field checking are marked with crosses. The obvious areas of rock outcrop and of alluvial floodplain soils in Plate 7(a) were delineated first. Outcrop areas were seen to be characterized by positive relief, light tone and ledges along the trace of the strike, below which the slope steepens and shadows impart a darker grey tone; alluvial soils were confined to narrow, flat, low-lying areas flanking the rivers and separated from the adjoining ground by marked changes of slope. Points for field checking were selected so as

(i) to sample each of these obvious units;
(ii) to investigate possible differences in the remaining areas which appeared to be associated with specific landforms such as colluvial fans and terraces;
(iii) to sample areas with different tonal expression on the aerial photographs and which appeared to conform to the orientation of the strike of the obvious outcrops; it was likely that these were due to differences in the underlying geology and consequently represented differences in residual soils.

As far as possible the points were located near to roads or footpaths to facilitate field access, while ensuring that they adequately covered the feature concerned.

The interpreter has now recorded all he can without field check, and his first airphoto interpretation has given him an overview of the area. He is now ready for his first field visit to familiarize himself with the area in detail. He should visit all the points selected during the initial airphoto interpretation and traverse all roads and tracks to look for exposures that may be examined. He will identify the soil units within each mapping unit and he will attempt to record a full soil profile (Chapter 6) within typical occurrences of each, usually from test pits or hand-auger holes of limited depth (1 to 2 m) situated at intervals of one or two kilometres along the route; more frequent intervals may be required in complex terrain or where natural exposures are sparse. A limited but representative number of soil samples for indicator tests will be collected from any soil horizons that are likely to present problems (e.g. compressible soils, active clayey soils, loose sands, etc.) or which may provide sources of construction material (e.g. gravels, sands, pedogenic materials, etc.). The positions of recorded soil profiles and sampling points are marked on the aerial photographs.

This preliminary set of soil profile data will form the basis for a provisional list of profile types. Once the list is complete, the aerial photographs of the observation sites are re-examined to determine the stereoscopic appearance of areas typical of each profile type, and a list of mapping units (areas which contain a uniform sequence of soil horizons) is compiled. An attempt is made to delineate the extent of each mapping unit by the interpretation techniques described in Appendix C. It will be found that some units can be mapped to contain only one profile type; others will contain two or more. The legend, of simple or compound units, is finalized. Only the central portions of each photograph should be annotated to minimize the inherent distortions in the photographic image. Indistinct or questionable boundaries between mapping units are shown as broken lines and will require further field checking. Although the majority of the units now tentatively mapped will have been sampled in the course of the visits which were planned during the first airphoto interpretation, some boundaries may still require verification. In the area of Plate 7(b) these were chiefly between units distinguished on the basis of vague and gradational tonal differences, the presence of which is related to subsoil variations, or which were obscured by cultural or agricultural features or veld fires. The boundary arrowed in Plate 7(b) was removed after field checking had identified the area of dark tone to be the result of a veld fire.

By this time the laboratory test data should be available and will indicate which soil types may present subgrade problems or offer potential sources of construction materials and therefore require further investigation. Armed with this information, the surveyor embarks on a second field visit to confirm the provisionally interpreted boundaries of the mapping units, devoting particular attention to those boundaries which proved difficult to recognize in the interpretation of the aerial photographs, and to identify and delineate those boundaries which could not be recognized on the photographs. He will also pay attention to variations that the indicator test data showed to be particularly important, e.g.

between clayey soils of low and high activity. He may require further test pits or auger holes to check boundaries and to investigate areas of problematical subgrade. Any units which may produce construction materials must be investigated in more detail. In this example (Plate 7(c)) the search for sources of fresh rock suitable for use as aggregates in concrete and bituminous surfacings was restricted to the obvious quartzite outcrops identified during the course of the initial interpretation of aerial photographs. Attention was given to local variations in quality as well as to access and workability. Two borrow areas of natural gravel for base materials were located where the lateral variability of the gravel was smallest and the thickness greatest, while the access and haul distance did not cause insurmountable difficulties. Similar considerations were applied in locating the other borrow areas.

Following this second field visit, the mapping units are finally delineated with minor modifications to the boundaries drawn earlier, and the proposed borrow areas and sources of water are added. All boundaries of mapping units may now be transferred to the photo-mosaic to ensure that there are no discrepancies at the edges of adjoining airphotos. The map units produced are very similar to the land facets of Chapter 10.

MAP PRESENTATION

Large scale topographic maps may be used as the base for the soil engineering map. The scale of this base map should be as close as possible to the scale of the aerial photographs; it should certainly not be larger. If no suitable topographic maps are available the base map may be compiled photogrammetrically or, if a lesser degree of accuracy is acceptable, the annotations may be directly traced off the airphoto mosaic or printed upon it.

The final form of the map will depend on the purpose to which it is to be put and on the preference of the users. A recommended form of presentation distinguishes all mapping units associated with the same bedrock by the same colour, together with a letter or two designating the stratigraphic unit. A dark shade represents solid outcrops of the particular rock type and a lighter shade of the same colour indicates where this rock is concealed beneath a mantle of soil. The locations of the test pits and the potential borrow areas should be shown.

Within a given colour, the soil profile typical of each mapping unit is represented by hatchings to indicate its predominant soil types, and by letters to indicate the different soil horizons and their sequence. These letters are usually abbreviations of the origins of different soil horizons, e.g. W/A/Rm/M represents windblown sand on alluvium overlying residual mudstone and finally mudstone bedrock. The map legend explains the use of the various symbols and abbreviations and relates them to various soil origins and their textures. Fig. 9.1 shows an alternative method of presentation without the use of colour.

The report which accompanies the map should include a typical soil profile for each mapping unit together with a summary of the laboratory test results for significant horizons in each mapping unit (Table 9.1).

SAMPLING AND TESTING FOR DESIGN PURPOSES

The soil engineering map and report provide a framework within which to carry out the detailed sampling and testing required for final engineering design. For some purposes the map units provide sufficient information in themselves—for others they indicate where further tests and field checks are necessary. The design engineer is also responsible for the investigation of potential sources of construction material; this will require closely spaced grids of test pits (at about 30 m intervals and about 2 m deep) and more rigorous laboratory testing including compaction and strength tests (Chapter 4). The engineer will also undertake a survey along the centreline, using the map as a guide to sampling intervals, in order to determine the range of properties of the subgrade materials occurring within each mapping unit. Morse (1961), Kantey and Williams (1962), and others have shown that the engineering properties of soils can be validly characterized in terms of such mapping units. One of their major conclusions was that sometimes only five samples were sufficient to define the index properties with the following coefficients of variation which were considered acceptable:

liquid limit $\pm$ 15 per cent of mean
plasticity index $\pm$ 25 per cent of mean
clay content $\pm$ 20 per cent of mean

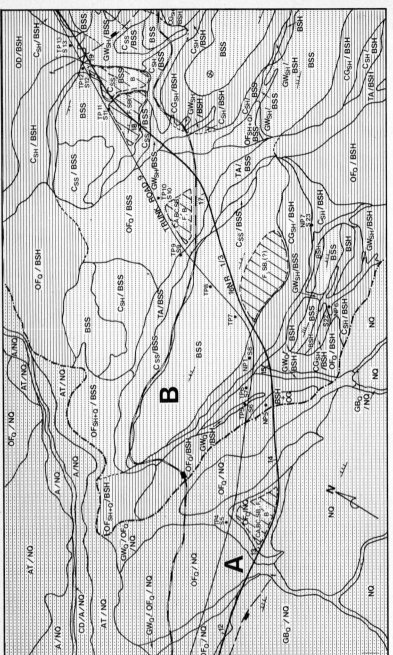

FIG. 9.1. Final soil engineering map for a major rural road. The area has been divided into two major terrain units (land systems, Chapters 10 and 11). Bedrock types are usually distinguished by different colours.

0 200 400 600 800   1   2

METRES   KILOMETRES

MAJOR TERRAIN UNITS AND MATERIALS SOURCES

GEOLOGY

B.S.S.	SANDSTONE AND QUARTZITE
BSH	SHALE
NQ	ORTHOQUARTZITE

LAND SYSTEM AND FACET

B, 1a, b, c.

B, 2a, b.

A, 1a, b, c (1).

LAND SYSTEM BOUNDARY AND SYMBOL:

A   TABLE MOUNTAIN QUARTZITE, SANDSTONE AND SHALE 368 LAND SYSTEM
B   BOKKEVELD SHALE AND SANDSTONE 365 LAND PATTERN

POSSIBLE SOURCE OF CONSTRUCTION MATERIAL:
CA = CONCRETE AGGREGATE; BC = BASE; SB = SUB-BASE;
CS = CONCRETE SAND; B = BINDER; F = FILL.

•TP2   TEST PIT LOCATION AND NUMBER

•NP3   NATURAL OR PREVIOUSLY EXCAVATED PROFILE LOCATION AND NUMBER

•S4   SAMPLE NUMBER

⊢—◢—f   FAULT INDICATING DOWNTHROW BLOCK

⊕   HORIZONTAL STRATA

⊥ 1°-10°   ⊥ 11°-25°   ⊥ 26°-45°   ⊥ <45°

TRANSPORTED SOILS

	ORIGIN AND COMPOSITION	UNIFIED CLASS	PROBABLE THICKNESS RANGE Metres	ASSOCIATED HYDROLOGY	LAND SYSTEM AND FACET
C_{SS}	COLLUVIUM OF SILTY FINE SAND DERIVED FROM SANDSTONE	SP-SM	0.5-1.5	OCCASIONAL SEEPAGE AND PERCHED WATER TABLES	B3
C_G	SCATTERED, POORLY SORTED SUBROUNDED QUARTZITE GRAVEL	GP	SUPERFICIAL	ABOVE GROUNDWATER INFLUENCE	B2b
CG_{SH}	ANGULAR MEDIUM AND FINE COLLUVIAL GRAVEL DERIVED FROM SHALE IN SILTY MATRIX	GM-GC	0.5-1.5	ABOVE GROUNDWATER INFLUENCE	B4
C_{SH}	COLLUVIUM OF CLAYEY SAND AND OCCASIONAL GRAVEL DERIVED FROM SHALE	SC	0.5-2.0	WATER TABLE AT MODERATE DEPTH	B5 (1),(2).
GW_{SS}	GULLY WASH OF SILT AND SAND DERIVED FROM SANDSTONE	SM-SC	1.0-3.0	PERIODICALLY INUNDATED	B6
GW_{SH}	GULLY WASH OF CLAYEY SAND AND OCCASIONAL GRAVEL DERIVED FROM SHALE	SC	1.0 - 3.0	PERIODICALLY INUNDATED	B7,(1),(2)
OF_{SH}	OUTWASH FAN OF SUBROUNDED GRAVEL AND BOULDERS IN WELL GRADED MATRIX	GC + BOL	3.0-10.0	PERIODICALLY INUNDATED	B8,(1),(2)
OD	OUTWASH DEPOSIT OF CLAYEY SAND AND OCCASIONAL GRAVEL	SC	2.0-4.0	PERIODICALLY INUNDATED	B9
TA	ALLUVIUM OF SILTY SAND OVERLYING SUBROUNDED GRAVEL AND BOULDERS IN A WELL GRADED MATRIX	GM/GC + BOL	2.0-6.0	PERIODICALLY INUNDATED	B10
T_G	TALUS OF POORLY SORTED, SUBANGULAR QUARTZITE GRAVEL AND BOULDERS	GP + BOL	1.0-5.0	OCCASIONAL SEEPAGE	A3 (1),(2),(3)
BF_O	BOULDER FAN OF POORLY SORTED, SUBANGULAR QUARTZITE GRAVEL AND BOULDERS	GP + BOL	2.0 - 10.0	OCCASIONAL SEEPAGE	A4 (1),(2),(3)
GB_O	GULLY FILLING OF POORLY SORTED, SUBROUNDED QUARTZITE GRAVEL AND BOULDERS	GP + BOL	2.0-5.0	PERIODICALLY INUNDATED	A5 (1),(2),(3)
OF_O	OUTWASH FAN OF SUBROUNDED QUARTZITE GRAVEL AND BOULDERS IN PREDOMINANTLY SANDY MATRIX	GW - GC + BOL	10.0-50.0	PERIODICALLY INUNDATED	A6 (1),(2),(3). B N C.1
GW_O	GULLY WASH OF SUBROUNDED QUARTZITE GRAVEL AND BOULDERS, OFTEN OVERLAIN BY THIN POORLY GRADED SAND	SP - SM/GW - GC + BOL	3.0-10.0	PERIODICALLY INUNDATED	A7 (1),(2),(3); B N C.2
AT	TERRACE ALLUVIUM OF INTERBEDDED CLAYEY, SILTY AND POORLY GRADED SAND	SP - SM - SC	5.0-15.0	HIGH WATER TABLE	A8 (1),(2),(3)
A	ALLUVIUM OF INTERBEDDED SILTY AND CLAYEY SAND	SM - SC	5.0-20.0	HIGH WATER TABLE	A9 (1),(2),(3)
CD	CHANNEL DEPOSIT OF POORLY GRADED, SUBROUNDED GRAVEL AND SAND	GP	2.0-5.0	PERMANENTLY INUNDATED	A10 (1),(2),(3)

TABLE 9.1

Summary of the engineering properties, and potential uses of materials, for land system A in Fig. 9.1

Facet and variant	Material	Profile Nos.	Sample Nos.	Depth (m)	Probable thickness range (m)	Unified class
1a† 1b 1c	Soft to very hard rock quartzite and sandstone	—	—	—	—	—
3(1)†, 3(2), 3(3)	Colluvial coarse sub-angular gravels and boulders	—	—	—	1.0–5.0	—
4(1), 4(2), 4(3)	Colluvial angular coarse gravel and boulders	TP16	19 20	0.8 1.3	2.0–10.0	SC SC
5(1), 5(2), 5(3)	Gully lining: sub-rounded gravel and boulders	—	—	—	2.0–5.0	—
6(1), 6(2), 6(3)	Sub-rounded gravels and boulders in a predominantly sandy matrix	TP1 NP2	1 15	1.0 0.5	10.0–50.0	GW GW
7(1), 7(2), 7(3)	As above, often overlain by thin poorly graded sand	TP2 TP4 NP6	2 5 16	1.0 0.6 0.8	3.0–10.0	SP-SM SC SP-SM
8(1), 8(2), 8(3)	Interbedded alluvial clayey, silty and poorly graded sand	TP3 TP3 NP4	3 4 21	0.3 0.6 3.6	5.0–15.0	SP-SM SC SM
9(1), 9(2), 9(3)	Interbedded alluvial silty and clayey sand	TP15 TP15	17 18	0.6 1.0	5.0–20.0	SM SM-SC
10(1), 10(2), 10(3)	Poorly graded sub-rounded alluvial gravels and sands	—	—	—	2.0–5.0	—

Facet 2 does not occur in the mapping area.
† a, b, and c indicate different outcrop landforms and (1), (2) and (3) are facet variants (Chapter 10).

PRA class	Activity	Laboratory testing	Base	Subbase
—	—	—	Suitable if fine-grained fresh, hard rock and crushed	Suitable if crushed, screened and binder added
—	—	—	As above but inaccessible	As above but inaccessible
A-2-6(2)	Medium	Indicator	Suitable if crushed	Suitable if crushed and screened
A-6(4)	Medium	Indicator		
—	—	—	As above but material limited and difficult to work	Suitable if crushed but material limited and difficult to work
A-1-a(0)	Low	Indicator + CBR	Suitable if crushed and screened	Suitable if crushed and screened
A-1-a(0)	Low	Indicator + CBR		
A-1-b(0)	Low	Indicator + CBR	May be suitable if crushed and screened	Suitable if screened
A-2-6(2)	Medium	Indicator		
A-1-b(0)	Low	Indicator		
A-1-b(0)	Low	Indicator		
A-2-6(2)	Medium	Indicator	—	—
A-2-4(0)	Low	Indicator		
A-2-4(0)	Low	Indicator		
A-4(0)	Low	Indicator	—	—
—	—	—	—	—

The number of samples from each map unit need not depend on its area, but it may be useful to adjust the number of samples from each unit to its variability. Morse and Thornburn (1961) have estimated that the number of samples required to define the properties of materials of different origin within these limits are:

> loess: 5 samples
> till: 9 samples
> glacial outwash and heterogenous alluvium: 12 samples (which highlights the difficulty of predicting the properties of these materials).

Potential sources of construction materials, or *borrow areas*, should ideally be located at intervals of about 3 km along the route to provide material for use as subbase, and at intervals of about 6 km for base (Chapter 4); however, the actual spacing will be dictated by the occurrence of potentially suitable materials. Ideally, fill is obtained from cuts and the choice of centreline attempts to balance cut and fill; if additional sources of fill material must be located these are sought as close as possible to their area of use. Potential borrow sources for any use must be investigated by further pitting and sampling at specified intervals; Table 9.2 illustrates such a specification. Test pits may have to be deeper than the initial exploratory pits in order to assess the full thickness of workable material, but are not usually taken below 3 m. Samples for index testing are taken from individual horizons, but *channel samples*, which combine all usable materials of the soil profile in their naturally occurring proportions, are usually collected for CBR testing.

Borrow areas smaller than about 1 ha in size or which will produce less than 20 000 m³ of material are likely to be uneconomic and should be avoided if possible. Table 9.3 provides a guide to the field appraisal of natural construction materials.

In areas of deep cut, or where bridges and tunnels must be constructed, the design engineer may call for geophysical surveys to show the configuration of the bedrock surface (Appendix D).

ABBREVIATED PROCEDURES

This method for preparing soil engineering maps is good general practice but may have to be simplified for low cost roads where the

TABLE 9.2

Typical specification for prospecting for road construction materials: the spacings of test pits and sampling intervals should be used as a rough guide only and may be varied in the light of the variability of the materials present in the borrow area

Proposed use	Spacing of test pits	Sampling intervals for index testing	Sampling intervals for CBR testing
Fill	120 m	One sample from each horizon thicker than 300 mm in each pit	One channel sample through all usable horizons for every 10 000 m^3 of material
Controlled layers (i.e. subbase and base)	60 m	One sample from each horizon thicker than 200 mm in each pit	One channel sample through all usable horizons for every 500 m^3 of material. In addition at least one sample should be taken from every 40 000 m^3 of material to determine the CBR using various stabilizing additives

main consideration is to locate suitable construction materials and problem areas. Inevitably a simpler procedure gives less information and depends more heavily on personal judgement. It may be possible to achieve some simplification by training less skilled airphoto interpreters to recognize the airphoto patterns of areas which have already presented difficulties, or have already yielded construction materials, and then to mark with crosses similar patterns elsewhere within the same general terrain type, for checking by field investigations. Under ideal conditions, particularly in arid areas, smaller scale photography (e.g. 1:100 000) may suffice for this purpose.

THE USE OF OTHER INFORMATION

The use of information contained on standard geological maps, surface drift maps, agricultural soil maps, and in geotechnical data banks has already been discussed in Chapter 8. Soil data are not required to as great a depth for planning roads, railways, canals,

TABLE 9.3

A guide to the field appraisal of potential road construction materials. (Modified from Weinert and Dehlen 1965)

Proposed use	Significant criteria	Quantity required for unit length of road	Production	Haulage and accessibility	Possible materials
Surfacing aggregate (chips)	Highest quality. Any fresh rock provided it is: (i) not laminated; (ii) not schistose; (iii) not too coarse-grained; (iv) containing little olivine, nepheline, muscovite For low polishability the rock should: (i) not have an homogenous mineral composition; (ii) not be fine-grained; (iii) not contain only minerals covering a small range of hardness	Small quantities only: <1000 m^3 km^{-1}. One big quarry can supply large areas: usually from the nearest commercial crusher	Quarry to considerable depth if necessary. Blasting and crushing required	As directed by economics: however large distances (even up to 100 km) permissible. Railing or road transport	Most igneous rocks; Quartzite; Hornfels; Tillite; Dolomite
Coarse aggregate for concrete	High quality but slightly lower than for surfacing aggregate. Criteria as above	Small quantities only	Quarry to depth if necessary. Blasting and crushing	Fairly long hauls permissible	As above, and quartzitic sandstone

Fine aggregate for concrete	Any pure sand and crusher dust of fresh, non-micaceous rock which is rich in quartz. Fresh rock containing olivine, nepheline, and sulphides (e.g. marcasite, pyrite) must not be used. Beware of shrinking materials, particularly argillaceous fragments	Small quantity	Crushing or borrow pits	As above, but usually locally available	Crusher dust from above. River sands. Washed residual granite or sandstone
Base	High quality. Disintegrated rock preferable to decomposed rock. <5% muscovite or sericite. Crushed rock required for major roads. Gravels may be suitable for minor roads—angular rather than rounded. Maximum size 75 mm	Medium quantity; 1000 to 5000 m^3 km^{-1}	Quarrying to considerable depth if necessary. Blasting, crushing. Ripping where disintegration prevails. Borrow pits for gravels at shallow depth. Gravels may be partly crushed	Fairly long distances: up to 25 km if nothing suitable nearby	All fresh igneous rocks. Arenaceous rocks. Sufficiently indurated argillaceous rocks. Tillite when passing hardness test. Alluvial and colluvial gravels. Calcrete and ferricrete
Binder for base	Fine-grained material of medium plasticity which should not contain active clays	Small quantity: about 1000 m^3 km^{-1}	Borrow pits to shallow depth	Short	Almost any medium plastic soil, e.g. fine colluvium, aeolian sands, residual acid igneous rocks

TABLE 9.3—(cont.)

Proposed use	Significant criteria	Quantity required for unit length of road	Production	Haulage and accessibility	Possible materials
Subbase	Medium quality. Quartzose soil or gravel. Avoid excessive montmorillonite (may stabilize with lime if montmorillonite cannot be avoided). Avoid excessive muscovite or sericite. Maximum size 75 mm unless can be broken by grid roller	Medium to large quantities; 2000 to 10 000 m³ km⁻¹	Borrow pits to shallow depth. Ripping. Shovelling	Medium to short haul	Nearly any weathered rock and soil with low plasticity, but avoid montmorillonite and sericite. Pedogenic materials. May be stabilized with lime or cement
Fill	Poor quality permissible. Avoid active clays. Better quality material required in high fills for stability	Large quantities: >10 000 m³ km⁻¹	Surface soils or materials from cuts. Ripping. Scraping or shovelling	Very short, and if at all possible, materials from the site	Almost any material that is not too plastic
Wearing course (gravel roads)	Very similar to subbase. Some plasticity necessary. Kaolinitic clay as binder, avoid montmorillonitic clay	Medium quantities: about 2000 to 4000 m³ km⁻¹	As for subbase	As for subbase	Any decomposed igneous rock of moderate plasticity. Clayey sand and gravel. Friable ferricrete and calcrete

TABLE 9.4

An example of the interpretation of agricultural soil data for engineering purposes. (After Helmer 1959)

Soil series and their underlying geological deposits	Soil association number	Soil association in mapped division No.	Appears on block diagram No.	OKLA. subgrade index No.	Texture	Permeability	Colour	PRA soil classification	Liquid limit	Plastic index	% passing No. 200 sieve	Calif. bearing ratio	% Asphalt	% Cement	Sub-grade Good	Sub-grade Fair	Sub-grade Poor	Vol. shrink. High	Vol. shrink. Med.	Vol. shrink. Low	Soil binder Yes	Soil binder No	Soil stab. Yes	Soil stab. No	Shoulder constr. Good	Shoulder constr. Fair	Shoulder constr. Poor	Erosion control Good	Erosion control Fair	Erosion control Poor	Seeding & sodding Good	Seeding & sodding Fair	Seeding & sodding Poor
Leflore	12-A	1,2	2																														
'A' 0–16"				25	M	VS	YB																										
'B' 16–40"				18	F	VS	YB																										
'C' 40–60"				25	F	VS	YB																										
'D' 60"+					F	VS	YB																										
Atoka formation								A-7-6(20)	60	31	94	4	No				×	×				×		×			×			×			×
McAloster formation								A-7-6(13)	45	22	92	4	No				×	×				×		×			×		×				×
Savannah formation								A-7-6(20)	63	32	98	4	No				×	×				×		×			×			×			×
Terrace deposits																																	
Lela	7	All	15,16,21,22																														
'A' 0–45"				16	F	VS	G	A-7-6(13)	42	20	90	5	No			×		×				×		×			×		×			×	
'AC' 45–75"					F	VS	RB																										
'C' 75–90"+					F	VS	RB																										
Alluvium				19	F	VS		A-7-6(15)	50	23	90	5	No			×		×				×		×			×			×		×	
Lightning	3(a)*	1,2,3,4,8	10																														
'A' 0–10"				11	MF	VS	G	A-4(8)	35	9	94	9	No			×			×			×		×		×				×		×	
'AC' 10–28"				11	F	VS	BG	A-4(8)	33	9	86	11	No			×		×				×		×			×			×		×	
'C' 28–40"+				18	F	VS	G	A-7-6(14)	47	23	94	4	No			×	×					×		×			×			×			×
Alluvium				19				A-7-6(15)	50	23	90	5	No			×	×					×		×			×			×		×	
Lincoln	6	All	11,15,18,23,25,26,29,30																														
'A' 0–15"				5	C	R	B	A-4(5)	24	4	61	4	No			×				×	×		×		×						×		×	
'C' 15–50"+				0	C	R	YB	A-2-4(0)	NP	NP	17	13	4.0			×				×	×		×		×						×			×
Alluvium				0				A-2-4(0)	19	2	21	15	4.0			×				×	×			×	×						×		×	
Linker	12(a)*	1,2	1,2																														
'A' 0–8"				0	MC	MR	G	A-2-4(0)	NP	NP	21	11	4.5			×				×		×	×			×				×		×	
'B' 8–40"					MF	MR	B																										
'C' 40–50"					MC	MR	R																										
'D' 50"+																																	
Atoka formation																																	
McAloster formation																																	
Savannah formation																																	
Stanley shale																																	

and pipelines as for urban planning, and so the relatively shallow depths to which soils are examined in agricultural soil surveys is less of a disadvantage and data from these maps may help greatly in the selection of mapping units and the early assessment of the engineering properties of their soils. An agricultural soil map is usually based on detailed fieldwork, so that the sequence of soil horizons to the depth which generally interests road design engineers, and the properties of each horizon, are relatively uniform within each series mapping unit (Table 9.4).

If a geotechnical data bank (Chapter 12) contains a large number of test results from an analogous area that may be linked to the mapping units within the area now to be mapped, this can be used to reduce the time and effort devoted to fieldwork and laboratory testing in the preparation of soil engineering maps of the new area. In this case the initial field visit is less vital, and the main effort is concentrated on delineating the mapping units and checking uncertain boundaries in the field.

ADVANTAGES AND LIMITATIONS

Soil engineering maps of this kind provide a rational basis for the rapid location of problem areas and construction materials and for selecting representative sampling points for the determination of subgrade conditions at appropriate intervals along the centreline. Dowling and Beaven (1969) have shown that such an approach may reduce field survey effort by as much as 70 per cent: in Fig. 9.1 the areas recommended as potential sources of construction material are confined to approximately 5 per cent of the total area with proportionate savings in the time required to identify specific borrow pits. The map facilitates communication between the contractor in the field and the design engineer in the office.

10. Terrain classification for engineering

Soil maps, resource maps, and engineers' planning maps implicitly assume that soil or terrain can be classified. By itself a description, or analysis, of the soil at a site only describes that site. Most engineering information is initially of this kind. However, if it were possible to classify all the terrain of an area into a manageable number of terrain classes—with all the terrain in one class reasonably similar (i.e. with each class uniform), and with all other classes different—then the information from any site could also be applied (with caution) to all other sites in the same terrain class. This will be essential if the engineer requires information on soil properties over a large area. Furthermore, such classes could be used to provide the framework for a data store—one 'pigeon-hole' to each class—or to index data in the store (Chapter 12).

If information collected at one site is to be used to plan operations at another in the same terrain class, then the classification must succeed in bringing similar sites into one class. As a corollary, when information is to be collected mainly at sites chosen because they are representative of a particular terrain class, the class must be sufficiently uniform to permit the meaningful extrapolation of this information to other sites in it. In order to classify the land over large areas in any reasonable time, remote sensing imagery—particularly aerial photographs which are readily available—may have to be used as the main tool for recognizing terrain class. The basic classes must, therefore, be recognizable on such images.

There have been two principal approaches to terrain classification—the parametric approach and the landscape or physiographic approach.

In its simplest form the *parametric approach* measures all relevant parameters of the terrain at a number of sites, defines terrain classes on particular ranges of the values of these parameters (e.g. a class defined on 5–10 per cent slope, CBR of 3–5 and clay content of 10–20 per cent in the surface soil horizon), and then interpolates the class boundaries between sites. In short, classes are defined on their engineering properties and then mapped.

The *landscape approach* recognizes terrain classes on their external features and their inter-relationships and then describes their properties from observations at representative sites in each, or by intelligent inferences from their genesis. For example, an alluvial fan is recognized on:

 (i) its occurrence at the foot of an escarpment;
 (ii) a lobate plan form with its apex at the mouth of a gully;
(iii) a convex cross-profile and a concave long-profile.

If resistant rocks such as quartzite are present in the escarpment, it is inferred, without a field visit, that the fan will contain gravelly colluvium, although confirmation is always desirable. In short, classes are defined because they are mappable and then their engineering properties are inferred or measured.

These two systems of terrain evaluation will be discussed in turn.

PARAMETRIC TERRAIN CLASSIFICATION

The landscape is observed at a network of sites, each of which is described in terms of its values of selected parameters, such as slope, drainage, and various soil properties. Values of the parameters are measured directly in the field (e.g. soil properties), from contour maps (e.g. slopes) or, less frequently, from remote sensing imagery (e.g. percentage vegetation cover), and the range of values is subdivided into specific classes (e.g. slopes of 0° to 2°, 2° to 5°, 5° to 10°, etc.). The number of parameters may have to be large if the purpose of the survey is general, but specific users may accept a classification of terrain in terms of only one or a few parameters.

Class intervals for each parameter are usually selected in relation to the specific constraints which they impose on different types of development, for example, gradients in excess of 5 per cent may be undesirable in a rural freeway, while the maximum permissible value for railways may be only 1.5 per cent. Where small terrain units are to be delineated at large scales (e.g. 1:20 000) it may be practicable to use several attributes in combination: thus a particular terrain unit may be defined in terms of characteristic slope, soil particle size distribution, soil moisture content and depth to bedrock.

Having decided which attributes and class intervals are relevant to the survey, an appropriate sampling procedure must be decided. Grid sampling facilitates computation and the interpolation of boundaries between terrain classes on the map, but stratified random methods, in which sampling sites are selected at random in areas which early observations show to be dissimilar, usually improve the reliability of the result. The sampling interval will depend on the map scale and the complexity of the terrain (Chapter 8). At larger scales it is possible to define classes which are more or less homogeneous in terms of a particular class interval; at small scales (1:50 000 to 1:100 000) the final mapping units will be defined in terms of the average or dominant value of a particular parameter, or, in complex terrain, by the sort of internal pattern which variations in a parameter display (e.g. the irregular changes of slope evident on recent glacial deposits): the latter method is a departure from the strict parametric approach as, like the landscape approach, it makes use of the recognition of landscape features to delineate boundaries.

Fig. 10.1 and 10.2 are examples of parametric terrain maps; Fig. 10.1 shows classes of plasticity within the surface soils of a relatively small area. Here the mapping units approximate in size to land facets (p. 214). Fig. 10.2 defines units of the same order of size in terms of ranges of slope.

Characteristic combinations of attributes may be recognized as different either by superimposing overlays of different parameters or by defining terrain classes in terms of different attribute values applied simultaneously, e.g. by means of cluster analysis (Webster 1977).

An advantage of the parametric approach is that arrays of field data, which have been systematically assembled for different parameters at a large number of closely-spaced sampling points, lend themselves to relatively simple statistical analysis and facilitate automatic interpolation and contouring by means of standard computer programs. Its main disadvantage, particularly when applied over large areas (e.g. for the planning of transportation routes), is the problem of locating the boundaries of parametric soil classes without a large number of field observations, and there are difficulties in establishing a flexible sampling system which takes into account local variations in the complexity of the terrain. While progress has been made in the automatic recognition and mapping of ranges of values of some surface terrain parameters, it is not yet possible to measure variations within the soil, particularly those which occur below the surface horizon. Thus while parametrically

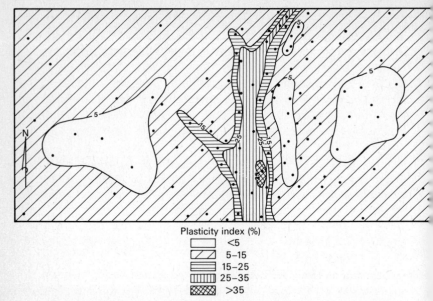

Plasticity index (%)
<5
5–15
15–25
25–35
>35

FIG. 10.1. The plasticity of the surface soils (0 to 300 mm) was mapped by a stratified random sampling system, with a greater sampling density near the central river channel in view of its greater complexity.

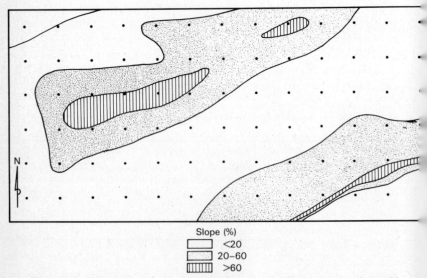

Slope (%)
<20
20–60
>60

FIG. 10.2. The slope of the land surface was mapped by regular grid sampling.

defined terrain classes have the potentiality for providing precise information, it often requires uneconomic effort to establish a sufficiently comprehensive data base to permit the accurate interpolation of boundaries.

LANDSCAPE APPROACH

In contrast, the landscape approach *starts* by identifying all obvious discontinuities in the landscape, either by field observation or using remote sensing imagery, and *then* undertakes a limited ground check to ascertain which tentative units to retain because they are different, which to combine because they are similar and which to subdivide by more laborious means because the areas within their boundaries are not uniform.

Unstead (1933) and Linton (1951) had already published some fundamental precepts of the landscape approach, but this was first applied systematically by Christian and Stewart (1953). The aim of their system of terrain classification was to provide a framework for a variety of data on natural resources accumulated during reconnaissance surveys of large areas. They discovered that, in many of their areas in tropical Australia, units of terrain defined in terms of their surface features and airphoto pattern could be used as a basis for extrapolating information from one place to another. A procedure soon evolved whereby detailed ground examination was restricted to small representative sampling areas and the units thus characterized were then identified in other areas of similar terrain and delineated by airphoto interpretation and a quick check. In this way the properties of extensive tracts of land could be quickly inferred and indexed at relatively low cost through the recognition of units analogous to those studied in the key areas.

Two different methods based on the landscape approach have been developed for engineering purposes during the last fifteen years, namely the land system/land facet classification developed in Oxford (Brink, Mabbutt, Webster, and Beckett 1966) and used in slightly modified form by the NIRR-Witwatersrand Group, in South Africa (Brink, Partridge, Webster, and Williams 1968), and the PUCE classification developed by the CSIRO in Australia (Aitchison and Grant 1967). Both are adapted from the approach originally developed by Christian and Stewart and both make extensive use of airphoto interpretation in recognizing and delineating terrain units over large areas.

Land system/land facet classification

The basic unit of this classification is the *land facet*, which is an area of ground with a simple surface form, a specific succession of soil profile horizons (each with reasonably uniform properties) and a characteristic groundwater regime. In undisturbed areas the land facet is also characterized by a locally distinctive plant association. A land facet may be delineated on aerial photographs at scales between 1:10 000 and 1:50 000, although in arid areas it may be possible to achieve delineation on scales as small as 1:80 000. Characteristically, land facets are small units and usually correspond to small physiographic features such as outcrops and free rock faces, talus slopes, alluvial fans, and alluvial terraces. There will be a large number of different facets within the boundaries of a single country, and so related facets are grouped into larger terrain units for convenience.

A recurrent pattern of genetically linked land facets is known as a *land system* (or land pattern). Usually each land system is dominated by one major geomorphic process. Land systems may be conveniently mapped at scales of 1:250 000 to 1:1 000 000. While they may be mapped from mosaics of small scale aerial photographs, satellite imagery has greatly facilitated the process (Fig. 11.2 and Plate 10).

A land system contains a characteristic group of facets recurring in a characteristic pattern, some of which may also be common to adjacent land systems. The boundary between an established and a new land system is recognized when:

 (i) the inter-relationship between land facets changes; or
(ii) the relative sizes of the land facets change significantly; or
(iii) a new land facet or group of land facets occurs in the landscape.

For example, one land system, representing a relatively mature stage of landscape development, may be characterized by a substantial proportion of inselbergs or upland residuals surrounded by small pediment aprons. Another, in a much more advanced stage in the cycle, may consist almost exclusively of coalescing pediments with a few steep residuals. Such distinctions between land systems are relatively easy to make and correspond to substantial differences in the proportions of contrasting slopes, rockiness, and depth of mantle. Similarly, it is useful to separate one land system, where a cliff and talus slopes intervene between the crestal areas and pediments, from another where the crests and pediments merge gently without the intervening facets. The marked change in relief will be significant for most purposes. These differences are illustrated in Fig. 10.3. The character of a land system and the interrelation between its constituent land facets can be usefully illustrated in an idealized block diagram. Thus Plate 1 is an annotated stereogram of a representative key area of a land system and Fig. 10.4 is the corresponding block diagram, while Table 10.1 gives the basic data for the definition of the constituent land facets. Such a table has become known as a *facet-index* and serves as a convenient summary of the properties of the facets; it should not be confused with the various types of index developed to provide access to data in a geotechnical data store. Note how the characteristic pattern of this land system derives from the weathering and erosion of interbedded hard and soft rocks and their dissection by small rivers. Plate 1 shows how the same facets recur in the same pattern many times over the entire width of the stereopair (more

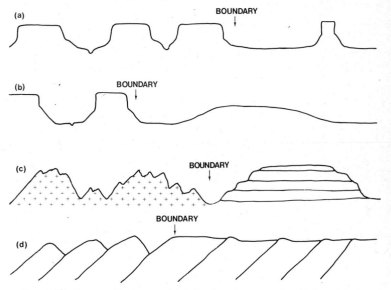

Fig. 10.3. Some criteria on which land systems can be distinguished.

than 5 km). Fig. 10.5 shows typical soil profiles for each land facet and variant. Once the land systems of an area have been delimited and each illustrated by a block diagram and facet definitions, the recognition of land facets is usually simple.

Underlying the definitions of these units is the idea of prediction: that by knowing and being able to recognize a land system one may predict, not only from one occurrence of it to another, but also about the parts of it from an understanding of their relation to the unit as a whole. One of the most important criteria for judging whether a land system or a land facet is satisfactory is that it can provide a basis for prediction. A satisfactory terrain classification of this type appears to require that the terrain of an area fits into a simple genetic picture. Where the genetic picture is complex, for example in certain areas of emergent coastline, glaciated shields and some till landscapes, it is easy enough to distinguish land systems, but it may be difficult to define and recognize facets within them (Lawrance, Webster, Beckett, Bibby, and Hudson 1977).

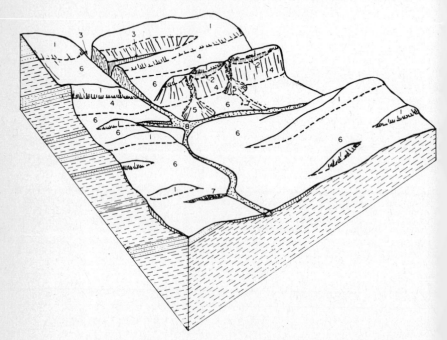

FIG. 10.4. Block diagram for the land system exemplified in Plate 1.

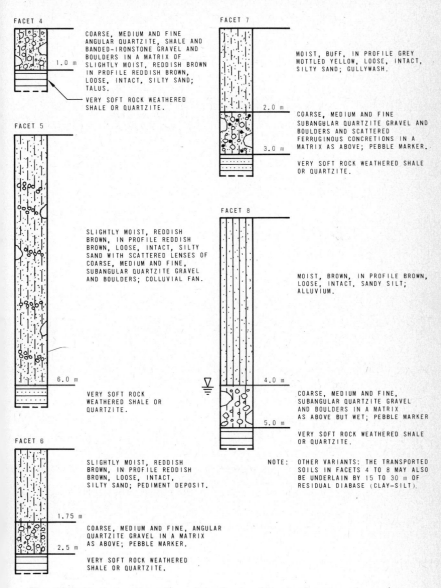

FACET 4

1.0 m

COARSE, MEDIUM AND FINE
ANGULAR QUARTZITE, SHALE AND
BANDED-IRONSTONE GRAVEL AND
BOULDERS IN A MATRIX OF
SLIGHTLY MOIST, REDDISH BROWN
IN PROFILE REDDISH BROWN,
LOOSE, INTACT, SILTY SAND;
TALUS.

VERY SOFT ROCK WEATHERED
SHALE OR QUARTZITE.

FACET 5

6.0 m

SLIGHTLY MOIST, REDDISH
BROWN, IN PROFILE REDDISH
BROWN, LOOSE, INTACT, SILTY
SAND WITH SCATTERED LENSES OF
COARSE, MEDIUM AND FINE,
SUBANGULAR QUARTZITE GRAVEL
AND BOULDERS; COLLUVIAL FAN.

VERY SOFT ROCK
WEATHERED SHALE OR
QUARTZITE.

FACET 6

1.75 m

2.5 m

SLIGHTLY MOIST, REDDISH
BROWN, IN PROFILE REDDISH
BROWN, LOOSE, INTACT,
SILTY SAND; PEDIMENT DEPOSIT.

COARSE, MEDIUM AND FINE, ANGULAR
QUARTZITE GRAVEL IN A MATRIX
AS ABOVE; PEBBLE MARKER.

VERY SOFT ROCK WEATHERED
SHALE OR QUARTZITE.

FACET 7

2.0 m

3.0 m

MOIST, BUFF, IN PROFILE GREY
MOTTLED YELLOW, LOOSE, INTACT,
SILTY SAND; GULLYWASH.

COARSE, MEDIUM AND FINE
SUBANGULAR QUARTZITE GRAVEL AND
BOULDERS AND SCATTERED
FERRUGINOUS CONCRETIONS IN A
MATRIX AS ABOVE; PEBBLE MARKER.

VERY SOFT ROCK WEATHERED SHALE
OR QUARTZITE.

FACET 8

4.0 m

5.0 m

MOIST, BROWN, IN PROFILE BROWN,
LOOSE, INTACT, SANDY SILT;
ALLUVIUM.

COARSE, MEDIUM AND FINE,
SUBANGULAR QUARTZITE GRAVEL
AND BOULDERS IN A MATRIX
AS ABOVE BUT WET; PEBBLE MARKER

VERY SOFT ROCK WEATHERED SHALE
OR QUARTZITE.

NOTE: OTHER VARIANTS: THE TRANSPORTED
SOILS IN FACETS 4 TO 8 MAY ALSO
BE UNDERLAIN BY 15 TO 30 m OF
RESIDUAL DIABASE (CLAY-SILT).

FIG. 10.5. Typical soil profiles of the facets of the land system shown in
Plate 1 and Fig. 10.4.

TABLE 10.1

The facet-index for the land system of Figs. 10.4, 10.5 and Plate 1

Land facet	Form	Soils, materials, and hydrology	Land cover†
1	Waxing slope or convex outcrop. Slope c. 0–45° width, c. 20–750 m	Fresh quartzite or shale outcrop. Above groundwater influence	Trees and shrubs *Protea caffra, Protea roupelliae, Bequaertiodendron magalismontanum, Iandolphia capensis, Tapiphyllum parvifolium* Grass. *Aristida* sp. Urban development
2	Free face. Slope c. 45–90°, height c. 5–15 m	Fresh quartzite face. Above groundwater influence	Trees and shrubs as above Grass as above
3	Rockbound gully or gorge. Slope c. 10–60°, width c. 50–400 m	Fresh quartzite or shale slope. Periodically inundated	Trees and shrubs as above Grass as above
4	Talus slope. Slope c. 15–35°, width c. 100–300 m	Talus and colluvial gravel (c. 1–2 m) of quartzite and shale on weathered shale or quartzite. Above groundwater influence	Trees and shrubs *Protea caffra, Protea roupelliae* Grass *Themeda triandra* Herbs *Paranari capense*

5	Fan. Slope *c.* 5–15°, width *c.* 100–300 m	Sandy and bouldery colluvium (*c.* 3–10 m) on weathered shale or quartzite. Periodically inundated	Crops (chiefly vegetables), orchards (chiefly deciduous) plantations (chiefly gum), and urban development. Natural vegetation very disturbed
6	Pediment. Slope *c.* 2–12°, width *c.* 100–1000 m	Hillwash of silty sand derived from quartzite and shale on weathered shale or quartzite. Occasionally saturated	Grass *Themeda triandra, Tristachya rehmannii, Elyonurus argenteus*. Land use as above
7	Low angle gully. Slope *c.* 2–10°, width *c.* 100–250 m	Sandy gully wash (*c.* 1.5–3 m) derived chiefly from quartzite on weathered shale or quartzite. Periodically saturated	Grass *Cyperus spherocarpus*. Land use as above
8	Alluvial floodplain. Slope *c.* 1–5°, width *c.* 50–100 m	Silty alluvium (*c.* 3–6 m) on weathered shale or quartzite. High water table	Trees and shrubs *Leucosidea sericea* Herbs *Pteridium aquilinum* var. *capense, Artemisia afra*. Land use as above

† Dominant species are listed here, but in most cases the dominant vegetation class (e.g. bush or grass) and density of cover will suffice

Although land facets and land systems are the main classes of the classification system, the former may not be a sufficiently fine subdivision of the land for all engineering purposes. The scheme therefore provides for subdivision of the land facet into *land elements* which are equivalent to the 'sites' of Bourne (1931). A land element is part of a land facet with a uniform lithology, form, and soil profile. Examples are the upper and lower segments of an alluvial fan and the shallow, braided distributory channels which cross its surface; prospecting for coarse granular construction materials would usually be confined to the steep upper segment. Land elements can usually be recognized on aerial photographs of 1:10 000 to 1:50 000 scales, although they cannot always be mapped at these scales.

Variations within a land facet which are important, but which are not predictable from surface evidence or from their position in the land system, are termed *variants*; they commonly arise from differences of substrata which are significant for engineering purposes, such as soils residual from different rock types concealed beneath a uniform mantle of transported soil. Although their identification invariably requires field investigation, variants should still be listed in the facet-index.

PUCE classification

The PUCE (pattern–unit–component–evaluation) classification was developed by the Geomechanics Division of CSIRO in Australia; it provides an hierarchical terrain classification for engineering purposes of four levels, namely the province, pattern, unit, and component. It is designed to operate at all scales of concern to the engineer, and the classes are based on easily recognizable characteristics which facilitate their use as mapping units.

The *province* is an area of constant geology at group level and its boundaries are delineated either directly from existing geological maps or by airphoto interpretation and geological reconnaissance. Provinces may cover several hundreds of thousands of square kilometres and can be mapped at a scale of the order of 1:1 000 000. The *terrain pattern* is defined chiefly in terms of airphoto pattern and is delineated at scales of the order of 1:100 000. It is further defined by field sampling as an area within which a constant association of terrain units occurs. *Terrain units* are areas occupied by a single physiographic feature, that is a 'characteristic association of earthen materials with a characteristic [natural] vegetative cover' (Aitchison and Grant 1967). These characteristics are determined chiefly by field study. Terrain units may be mapped at scales of 1:10 000 to 1:50 000. A terrain unit is not uniquely associated with a specific terrain pattern. One terrain unit may in fact occur in several terrain patterns, and the latter may differ merely in changes in the proportions of their constituent terrain units. The *terrain component* is a unit characterized by a consistent soil profile and vegetation association and a constant rate of change of slope. Terrain components are not usually mapped and are not uniquely associated with any specific terrain unit.

Descriptive and quantitive data are incorporated into the classification as in Table 10.2. To employ this system the terrain *pattern*, the terrain *unit*, and the terrain *component* at the site of interest must be appreciated in turn and their recorded properties then evaluated for the purpose of the moment.

In general, the landscape approach exploits the repetitive nature of terrain and allows the classification and mapping of large areas relatively quickly and inexpensively by the interpretation of remote sensing imagery. The classification merely provides a framework and each class still needs to be filled with information. Inevitably, the extensive use of analogue matching, with limited sampling of restricted key areas and the extrapolation of these data to other areas by means of indirect identifications on aerial photographs or other images, involves some degree of generalization. But the experience of the interpreter will minimize these disadvantages, and they may be compensated by speed and low cost. This approach seems to provide the only practical method for the rapid acquisition of data with sufficient accuracy to permit planning over large areas at *project level*, i.e. for the location, design, and construction of a specific transportation line such as a road. Table 8.1 illustrates the kind of generalized information that can be made available: the materials listed in the table occur as individual horizons within facets or variants in a single land system.

These comparisons have stressed the differences between the main techniques of terrain classification but the different approaches are not always incompatible. Thus the low cost and

TABLE 10.2

Description of terrain classes in the PUCE system. (After Aitchison and Grant 1967)

Terrain class (1)	Terrain factors suitable for descriptive expression (2)	Terrain factors suitable for quantitative expression (3)	Methodology of quantification (4)	How to retrieve input in the class (5)
Terrain pattern	Geomorphology Basic characteristics of soil, rock, vegetation common among constituent terrain units Drainage pattern	Relief amplitudes Stream frequencies	Airphoto or ground study Airphoto study	Via terrain pattern map and legend
Terrain unit	Physiographic unit Principal characteristics of soil, rock, vegetation	Dimensions of physiographic unit (relief amplitude, length, width)	Airphoto or ground study	Via terrain pattern map and legend of constituent units plus recognition of physiographic characteristics of terrain unit or by direct recognition of physiographic and associated characteristics of terrain unit

Terrain component	Physiographic component Lithology Soil type Vegetation association	Dimensions of physiographic component (relief amplitude, length, width, slopes)	Measured *in situ*	Via recognition of terrain unit by either method as above plus recognition of physiographic and associated features of terrain component
		Dimensions of vegetation (height, diameter, spacing)	Measured *in situ*	
		Dimensions of surface obstacles including rock outcrops and termitaria	Measured *in situ*	
		Properties of earthen materials throughout profile: Depth Particle size gradation Consistence Strength Permeability Suction Mineralogy	Measured in the field where practicable; otherwise by standard laboratory procedures	
		Quantities of earthen materials	Measured or estimated *in situ*	

moderate reliability of the landscape approach, which are ideally suited to more extensive investigations, may be modified by more frequent sampling to suit the requirements of more intensive planning. Indeed, the distribution of the sampling points for a more detailed survey may be improved by a preliminary breakdown of the terrain into landscape units, while the later interpolation of parametric boundaries between observation points may be aided by observing airphoto boundaries between these points. Conversely, where boundaries between units are vague, intensive sampling may improve their accuracy and may increase their uniformity. Combinations like this are particularly useful in the more intensive soil investigations demanded in the planning of townships, large industrial developments, and in investigations for earth dams and airfields.

It will be useful to review the different approaches to soil survey described so far. Site investigations for individual structures (Chapter 7) rely almost exclusively on the parametric approach, while soil surveys of areas of intermediate size for urban planning purposes (Chapter 8) tend to define soil (terrain) units parametrically and then to use remote sensing imagery among the various aids to mapping them. In size, the mapping units correspond to land facet variants or even land elements. Chapter 9 is concerned with the mapping of land facets to guide the choice of transportation routes and the most economic deployment of a given sampling effort. Less frequently, facet variants are mapped where land facets must be subdivided further to define the extent of a specific problem area or source of construction material. Chapter 11 discusses land system mapping to select the best alignment for a transportation route and for regional planning.

11. Land system mapping—regional soil mapping for engineering

In surveys of large areas, small-scale geological and pedological maps may provide invaluable basic data which can sometimes be augmented and interpreted for engineering applications (Chapter 10). It is, however, often necessary to undertake a regional soil engineering survey *ab initio* to provide rapid information on the principal advantages and inherent constraints and distribution of soils and natural construction materials, in order to aid the location of physical development (e.g. the alignment of rural roads and other communication lines).

Small-scale surveys are usually multi-disciplinary and record other elements of the environment such as topography and vegetation. Maximum use must be made of remote sensing techniques to identify and delineate mapping units. Dowling and Beaven (1969) have pointed out that it is possible to make quite useful if broad generalizations about the suitability of individual land systems for road location, the availability of construction materials, suitable construction methods and costs of construction and maintenance. Table 11.1 shows how regional data of this kind may be interpreted for route location. Such general information is particularly useful in the early stages of planning in a developing country. In general, then, the landscape will be analysed in terms of a terrain classification. There will certainly not be enough field data, at the stage of planning we are now considering, to develop a parametric classification. Mapping units will usually be delineated on remote sensing imagery by the analogue approach (Chapters 8 and 9) with an irreducible number of field sampling points. Surveys of this type should make the maximum use of any relevant data in a terrain data bank (Chapter 12) from analogous areas.

BASIC ELEMENTS OF THE APPROACH

Surveys to assist the design of transportation routes (Chapter 9) required the delineation of land facets and facet variants as mapping units, but regional surveys are usually at smaller scales

TABLE 11.1

The land system provides a basis for systematic records of information about the engineering features of terrain (from MVEE)

Indexing Attributes:	Name DOTO		Abst. No.	Lf No.	
Clim.	Lith.	Soil	Vegtn.	Alt.	Relief
Morpho Tectonic	Divsn.	1:1 m			
Def. by	Date	Ref.	Org.		

Description of Land System as a whole.

[*Climate (station and data); lithology/stratigraphy; genetic links and geomorphic patterns; soil; vegetation; physiognomy; altitude; relief; location (1:1 m map sheet).*]

CLIMATE: Semi-arid Equatorial Tropical (Sudanian or Northern Savanna Belt). 1.532 (Papadakis), 3 humid months (July–September) 750–850 mm Rainfall. Average daily maximum temperature above 33.5 °C.

ROCK: Medium to coarse grained sandstones of the Palaeocene Kerri-Kerri Formation with subordinate conglomerates, clays and siltstones; flat lying strata with an overall gentle dip to the north. Overlain by a thick, extensive lateritic ironstone which dips beneath Quaternary sediments of the Chad Formation.

LANDSCAPE: Extensive dissection has given rise to an intricate system of deep, steep sided, flat bottomed valleys, with tabular interfluves and residual mesas, frequently capped with lateritic ironstone. The form of the drainage pattern shows that the former stream courses were part of a system which flowed N.E. to Lake Chad but has subsequently been captured by a tributary of the R. Gongola (R. Gaji) with a reversal of drainage. The accumulation of sediments on valley floors demonstrates that active down cutting has now ceased.

SOILS: Thin, stony soils on lateritic ironstone and sandstone. Sandy colluvium and alluvium of flat floored valley bottoms.

VEGETATION: Sudan Savanna zone; with mixed Detarium woodland on stony upper slopes. Savanna 'parkland' on cultivated valley floors.

ALTITUDE: 500 m (approx.)

RELIEF: 100 m.

REF: Klinkenberg, K et al. Soil Survey Section Bull. No. 21. 1963. Regional Research Station, Ministry of Agriculture Samara, N. Nigeria.

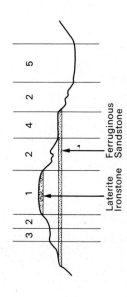

Laterite Ironstone

Ferruginous Sandstone

Add any comments necessary to clarify the diagram and to indicate the lake and inter-relations between all its facets use a second diagram if necessary to indicate the grain and texture of the country.

Land facet	Form	Soils, materials and hydrology	Land cover
1	**LATERITIC IRONSTONE SURFACE** Flat to gently sloping surface at 600 m. O.D. Occurrences vary in size from 100 m² to 2 km². Occurs in flat interfluves and isolated mesa surfaces	Lateritic ironstone, 3 in. thick, occurs as surface capping to sandstone, the upper part of which is often heavily ferruginised. Surface materials comprise bare ironstone and thin stony soils. Where the laterite has been stripped away, stony soils overlie sandstone	Mixed Detarium woodland where sufficient soil occurs
2	**SCARP SLOPE** Uneven steep topography generally becoming steeper on upper part of slope below margin of Land Facet 1. Concave lower slopes with uneven bouldery micro-relief. Steep, rocky ribs and pinnacles and incised gullies are common	Lateritic ironstone and sandstone rock, talus and thin, stony soils subject to erosion by hill wash	Detarium woodland where sufficient soil occurs
3	**SANDSTONE SURFACE** Flat to moderately sloping ground corresponding to a structural bench formed by dissection of flat-bedded sandstone	Rock and thin, stony soils	As above
4	**FERRUGINOUS SANDSTONE BENCH** As above, but more persistently developed and determined by the presence of a ferruginous sandstone horizon 0.5 m thick, midway between Land Facets 1 and 5	Rock and thin stony soils. The indurated, ferruginous sandstone protects the underlying softer sandstone and forms a prominent bench, often bare rock but sometimes overlain by rubbly soils	As above
5	**VALLEY FLOOR** Flat, but steepening slightly at margin with Land Facet 2. Vary in width from 10 to 900 m; Length may be as much as 13 m	Moderately deep, light grey to reddish brown sandy colluvial and alluvial soils. Permeable with no well-defined watercourses. Active downcutting of the gullies has now ceased	Savanna 'Parkland' heavily cultivated

Indicate the kind of predictable variation within facets, and between elements, if necessary by a sketch section

a. Engineering Properties of materials

Laboratory data on representative samples

Land facet	Dominant Engineering soil class (USCS)		Plasticity Characteristics		BS Compaction (light)		CBR at equivalent BS compaction (light) (per cent)	Particle size distribution (%)							Depth	Drainage (Internal)
								Gravel			Sand			Silt and Clay		
	Name	Symbol	Liquid Limit (per cent)	Plasticity Index (per cent)	Max. Dry Density (lb/ft³)	Opt. Moisture Content (per cent)		Coarse	Med	Fine	Coarse	Med	Fine			
1	Ironstone Rock Well graded gravel	GW	Non-plastic		138	10	36	5	20	35	10	10	8	12	Variable 5–15 ft	Generally well-drained. Short-lived local impedance due to lateritic caprock
	Exposed surfaces generally massive and indurated in contrast to the slightly softer material at depth. Hard, coarse, nodular and vesicular. Hand specimens of the nodules are frequently pisolitic, exhibiting dark coloured concentric iron concretions															
2	Comprises a wide range of engineering materials (sandstone rock, ill-sorted debris etc.) Pockets of finer material	GP GW	Sandstone rock Non-plastic 39	17				42 Nil	35 2	7 7	3 8	7 11	4 15	2 57		Well drained but subject to powerful erosion
3	Sand/silt. Sandstone rock	SM	Non-plastic		124	9	36	Nil	Nil	2	11	12	17	58	Variable 6 inch 3 ft	Generally well-drained, occasional perched water-tables form where clay-pans are present in the profile
4	Indurated rock		rock													
5	Silty sand and clayey sands with some sandstone fragments Poorly graded	SM/SC SP	Non-plastic 23–26 Non-plastic	6–11				Nil Nil	Nil Nil	5 Nil	23 15	32 45	15 32	25 8		Well-drained Well-drained but considerable surface sheet-wash in vicinity of scarp slopes

b. Engineering Resources

CONSTRUCTION MATERIALS

Land facet	Water Supply	Concrete Sand	Concrete Aggregate	Timber	Road Works			
					Sub base	Base	Chippings
1	Very feeble supply from shallow wells in dry season contaminated by surface flow in wet season		Indurated surface rock after crushing and screening		Suitable but uneconomic Suitable plentiful supply	Suitable, under caprock. Requires ripping and crushing Suitable but requires stabilization with lime or cement	Suitable, after screening of crusher run material Suitable after screening
2			Plentiful supply of medium-hard sandstone rock suitable for crushing		Gravels suitable but rock uneconomic	Rock suitable after crushing. Gravels suitable only when stabilized in cement or lime	After screening of crusher run material
3	Generally deep wells. Occasionally, perched water tables supply shallow wells				Suitable		
4			Indurated rock suitable for crushing			Suitable after crushing	Suitable after screening of crusher run materials
5	Perennial supply of potable water from deep wells. Liable to contamination by surface flow in wet season	Limited supply in local pockets deposited by sheet wash		Orchard bush. Suitable for kindling but not for construction	Suitable		

c. Suitability for Road Location

Land facet	APPRAISAL	FEATURES
1	Unsuitable due to isolated facet situation. Where same facet is extensively developed in adjacent land-system then suitable for road location	Flat to gently undulating topography. Abundant supplies of gravel. Generally well drained
2	Generally unsuitable because of steep slopes	Steep rock, with sliding and erosion
3	Suitable when extensively developed	Soils are sandy and well-drained with few well-defined water courses
4	Unsuitable because of isolated facet situation	Isolated situation and access is difficult. Dip of strata is favourable
5	Suitable .	Soils usually deep and well-drained. There are no excessive slopes. Locally run-off from steep slopes of land facet 2 may create an erosion problem and intercepting ditches and culverts would be required

and based on a smaller amount of field information, and so their mapping units are almost always land systems or their equivalent, e.g. the terrain patterns of the PUCE system. Representative key areas may be mapped in land facets to illustrate the range of variation within each land system and the manner in which its constituent units are inter-related. The detailed field sampling necessary to determine the engineering potentiality of each land system is usually also restricted to such key areas, and it has to be assumed that the properties thus measured in the facets and land systems also apply to all other examples of the same terrain units. This is, of course, the basis of the analogue approach, and is shown schematically in Fig. 11.1.

The map scale will generally fall in the range 1:250 000 to 1:1 000 000, but it may have to be increased to 1:50 000 in complex terrain.

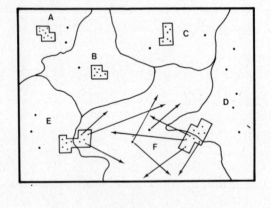

A–F   LAND SYSTEMS

REPRESENTATIVE AREAS MAPPED IN DETAIL AND SAMPLED

. . .   SAMPLE POINTS ON RECONNAISSANCE TRAVERSE

EXTRAPOLATION TO ANALOGOUS AREAS

FIG. 11.1. A land system map provides a basis for generalizing information and a framework for more precise generalization at facet level.

DETAILED PROCEDURES

The area to be investigated will consist either of a block of terrain or of a wide corridor, 10 to 50 km in width, within which the transportation line to link specified centres is to be located. Once the limits of the area have been defined the surveyor will assemble all available information from geological, topographic, or pedological maps, which in many cases may be of small scale (1:250 000 or smaller), and aerial photographs or other remote sensing imagery. In view of the small scale of the final map the imagery should be at a scale of 1:80 000 or smaller, and certainly no larger than 1:50 000. A mosaic is assembled for the corridor or area of interest, and a second set of loose prints of the same imagery is retained for interpretation.

The interpreter first makes a rapid scan of the imagery, with constant reference to the other information available and to the overall pattern displayed on the mosaic, and tentatively divides the whole into sub-areas of a size roughly corresponding to land systems (Chapter 10), each characterized by a particular type of terrain or a particular pattern of terrain types. At this stage the division is derived from general impressions rather than from rigorous criteria based on ground truth. Their broad view and lack of minor local detail make satellite images particularly useful at this preliminary or over-view stage. Even at this stage, any areas with a topography obviously unsuitable for route location may be excluded from further examination.

Having divided the whole area into tentative land system units, the interpreter examines the mosaic again and selects a representative key area for each land system which fully encompasses its whole range of conditions. Usually there is more than one occurrence of any land system or terrain type; the separate local occurrences should be compared carefully to make sure that the key area selected is fully representative of them all. If direct access to some parts of an area is not possible (as in some very rugged or remote terrain), tracts of analogous terrain should be sought in accessible areas in which the attributes which govern the formation and properties of soils (viz. bedrock geology, climate, relief, and stage of the geomorphic cycle) are the same as in the inaccessible areas. It is convenient if a key area is confined within one stereo-overlap (approximately 100 km^2) of the imagery being used. Having selected the key areas, the prints that cover them may be enlarged to two or three times the scale of the mosaic for more detailed analysis.

An initial stereoscopic study (Plate 2) should attempt to delineate units comparable in size to the land facets defined in Chapter 10. It is useful to prepare a tentative block diagram to conceptualize the geometric inter-relationship of the tentative land facets within each land system. Points which are easily accessible for field checking and appear to be representative of the soil conditions in each facet should be marked with crosses.

Each key area is then ground-checked (Chapter 9). If possible, two or three test pits or natural exposures should be examined in each facet in order to discern any variations within it (facet variants, see p. 220). Their soil profiles are described, and representative samples are taken of any materials which may present problems or which may be useful in construction. It may be useful to record the strength of the near-surface soil at different moisture contents by means of cone-penetrometer probes and to augment these field observations by wetting and testing samples of any soil which may possess a low shear strength when moist. Some of the tentative facet boundaries should be confirmed at points accessible to rapid traverses along existing roads or tracks or by helicopter, not only to confirm the distinctness and significance of the facets, but also to check their recognition features.

In the light of these point checks and the laboratory data a facet-index is compiled for each land system (Table 10.1, p. 218). The block diagram is checked and amended. A boundary is recognized between two land systems where a different assemblage of land facets is encountered, or where the size of one or more key facets changes significantly, or where the genetic relationship between various facets changes (Chapter 10). Then the land system delineations are checked and amended on the mosaic in the light of revised definitions and inventories. Changes in drainage patterns are particularly useful in identifying boundaries between land systems, and the synoptic view afforded by satellite images is very helpful in recognizing and delineating whole land systems (Plate 3).

MAP PRESENTATION

The mosaic annotated with land system boundaries is usually at a considerably larger scale than that at which final information can most conveniently be presented. So that its scale may be reduced by as much as five times when it is reproduced, it is essential that the land system boundaries be boldly drawn in a colour sensitive to photography (e.g. in red wax pencil). Since the mosaic will almost always be uncontrolled, there will be scale variations from place to place, particularly in areas of high relief and near the edges of individual images; this must be borne in mind if the annotations are to be transferred to an accurate base map (e.g. the 1:250 000 series available in many countries). Adjustments will have to be made to accommodate these distortions. Adjustments are facilitated by accentuating the drainage patterns on the mosaic and matching these with the corresponding patterns on the map.

In this type of map, simplicity of presentation is crucial. The areas occupied by different land systems may be distinguished by different colours or hatchings, but bold outlines and identifying numbers and letters which stand out against the background of the base map or of the mosaic often suffice (Fig. 11.2). The accompanying report will include facet-indexes, and block diagrams and representative annotated stereo-pairs of each land system. It will draw particular attention to the advantages and disadvantages of each unit for cross-country trafficability or for the construction of the proposed transportation line (Table 11.2), such as the proportions of each land system which are occupied by poor or potentially problematical subgrades, unstable slopes or periodically inundated areas, and good natural subgrades or potential sources of construction material. If the advantages of each land system can be compared in this way it is easier to select the best route from the geotechnical viewpoint. Zones of preferred terrain that emerge from this type of analysis can usually be linked to form one or more preferred corridors within which to locate the route line.

MERITS OF THE TECHNIQUE

Generalized terrain maps of this kind have been found useful for providing an inventory of construction materials, and of potential soil problems, rapidly and at relatively low cost for engineering projects in developing countries. Mapping at this level of generalization can be done directly on satellite imagery (Plate 3). Large-scale

TABLE 11.2

Summary of the terrain conditions, in terms of land systems, affecting the

Land system	Type of surface form	Major lithology	Approximate altitude range (m)	Approximate relief (m)	Route in relation to surface configuration and proposed changes
PS.272	Gently undulating with sandy valley floors	Lavas and shales with covering of aeolian sand in valleys	1225–1375	< 30	Crosses most facets No change
RS.261	Very gently undulating sandy surface with occasional shallow pans	Aeolian sand	1150–1375	6	Crosses most facets No change
MQ.276	Very steep folded quartzite ridges of high relief with narrow sand filled valleys	Quartzites	1150–1675	350	Entirely avoided No change
RS.161a	Parallel low sand dunes trending NW–SE, largely stabilized by vegetation	Aeolian sand	900–1225	6	Transverse to dunes No change
KS.174	Very steep folded schist and quartzite ridges projecting through gently undulating sand cover	Schist projecting through aeolian sand	1225–1375	120	Entirely avoided No change
RS.161b	Deranged barchan dunes, stabilized by vegetation	Aeolian sand	1150–1225	3–6	Avoided. No change
KS.153a	Fairly steep sub-parallel ridges following fold patterns, with sandy valley floors	Schist	850–1000	Approx. 50 but somewhat variable from place to place	Mainly along valleys No change
MQ.162	Gently undulating folded ridges with thin covering of largely stabilized parallel dunes trending approximately NW/SE	Meta-morphosed sediments	1000–1100	30(?) (Dunes ± 8)	Entirely avoided No change

selection of a route for a railway

Major soil types and approx. thicknesses	Gradient conditions	Extent of cut and type of excavation required	Foundation conditions	Other problems
Slightly weathered outcrop on hill crests and wind-blown sand approx. 5 m thick in lower areas	Gentle in all facets	Fairly shallow through hill crests, may require blasting	Adequate bearing capacity at surface except in non-cohesive sands	
Windblown sand up to 15 m thick	Very gentle in all facets	Little or no cut necessary, excavation with normal mechanical plant	Non-cohesive sands will require cover. Collapsing grain structure likely	
Fresh outcrop on ridges, windblown sand and hill-wash >15 m thick in valleys	Mostly very steep	Extensive tunnelling would be required	Non-cohesive sands along valleys would require cover. Collapsing grain structure likely	
Windblown sand at least 15 m thick	Very gentle except along flanks of dunes	Little cut required except to transect larger dunes. Excavation with normal mechanical plant	Non-cohesive sands will require cover and low batters in cuttings. Collapsing grain structure likely	Mobile sand area indicated on map, may encroach on line
Fresh outcrop on ridges, windblown sand and hill-wash >15 m thick in valleys	Gentle along valleys, but very steep on ridges	Little or no cut necessary in valleys, excavation with normal mechanical plant	Non-cohesive sands would require cover. Collapsing grain structure likely	
Windblown sand at least 15 m thick	Very gentle except along flanks of dunes	Little cut required except to transect larger dunes. Excavation with normal mechanical plant	Non-cohesive sands will require cover and low batters in cuttings. Collapsing grain structure likely	Mobile sand area indicated on map, may encroach on line
Foliated outcrop on ridges, sandy alluvium up to 6 m thick in valleys	Fairly gentle	Some cut necessary through ridges—will require blasting	Adequate bearing capacity in schists but alluvial sands will require cover	
Fresh outcrop on ridges, crossed by dunes of wind-blown sand, up to 8 m thick	Fairly gentle	Some cut would be necessary through ridges and sand dunes—former would require blasting	Adequate bearing capacity in sediments, but windblown sands will require cover	Possible mobile sands in dune field

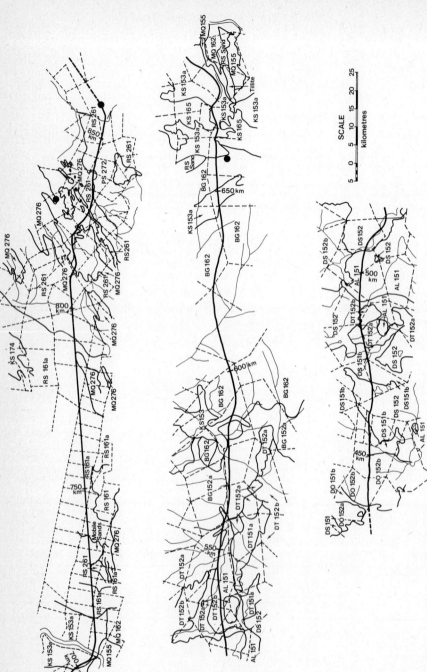

Fig. 11.2. A land system map to guide the location of a railway route (to be studied in conjunction with Table 11.2).

imagery is very useful for defining land facets in key areas but if necessary this can be done without remote sensing aids.

The principal advantage of this approach lies in its use of analogues: if any part of the terrain is analogous to that for which information has already been obtained elsewhere, it may be possible to extrapolate that information to the area now being mapped. A data bank will provide information accumulated from analogous areas in a simply retrievable form (Chapter 12).

Its principal drawback is that it places great reliance on indirectly acquired data, through the very extensive use of remote sensing imagery in the mapping process and through the widespread use of analogues in assigning properties to the soils within the mapping units. Inevitably the areas delineated at this small scale contain within their boundaries a wide range of soils with different engineering properties. The mapped land systems will be meaningful and useful in proportion as the soil units they contain are relatively uniform and lie in an easily recognizable pattern. If the component soil units of a land system can be seen to lie in a regular pattern it is often possible to predict their locations approximately, even without mapping them in detail.

12.   Geotechnical data banking

Over many years, measurements have been made at many sites in almost every part of the world. However, most of this information has been used only to plan construction at the sites from which it came. If it had been possible to match unsampled sites to sites where measurements have been made, the information from the latter could have been used at the former. Recognition of such similarities is a matter of personal intuition and memory; unfortunately these are limited, and so the application of prior knowledge to new sites has never been very extensive. As development accelerates it will be necessary to make wider use of existing information to avoid expensive duplication of effort, and it will be essential to develop systems to facilitate this.

The total amount of engineering and other terrain data generated within a national territory is likely to be immense. Tomlinson (1968) estimated that land data surveys of the developed areas of Canada would produce about 30 000 map sheets; there are likely to be some 20 m of boundary lines on any typical map sheet of 750 × 750 mm. Enormous files of data are necessary to characterize the mapping units delineated by these boundaries, and the individual records from which these characteristics were generalized would occupy even more space. It is clear that the sheer volume of fact demands the establishment of some kind of library with suitable indexing and retrieval mechanisms. If it is to be of real advantage a data storage and retrieval system must enable stored information to be applied to sites other than those from which it was gained. Further, if a system existed it might then be possible to plan the collection of new data with a view to both present and future projects: there would be a more efficient use of manpower if data were assembled and stored in a systematic way in contrast to the haphazard accumulation of small batches of unrelated information from separate development projects. It is, of course, inevitable that data collected systematically to cover very large areas will initially provide less precise information about any particular area than the data from a specific project, but as data accumulate it will become possible to provide increasingly specific data for substantial parts of a territory.

If information from sites sampled on one project, such as a road, is to be used to make tentative predictions about other but similar sites on a later project, then the system must include some means for matching sites. *Either* every site must be described in such detail (e.g. in terms of climate, geology, and physiography) that the soil at any other site described in the same way must be similar, *or* every site must be described by its class in a regional terrain classification. We have experience of a data bank, based on a landscape classification such as that described in Chapter 10, which is capable of storing and retrieving soil properties relevant to civil engineering applications. Lawrance (1972, 1975) has proposed similar ones for West Malaysia and for Ethiopia, and Grant and Lodwick (1968) describe one based on the Australian PUCE classification.

Depending on the level of accuracy demanded by a new project, and the reliability of the stored information, the output from a geotechnical data bank may be applied (i) directly to select the most favourable road alignment through land systems which have already been defined or (ii) to guide the selection of drilling equipment and the layout of trial holes for a programme of detailed site investigation appropriate to an individual structure. The somewhat tedious sequence of operations suggested for engineering soil surveys for roads and railways (Chapter 9) may be short-circuited if the data store already contains reliable information about the local land systems, together with annotated stereopairs of representative aerial photographs; this could reduce the need for an initial airphoto interpretation and a reconnaissance field visit. If so, it may be possible to produce an acceptable map of land facets on the imagery with only one field visit to identify facet variants, and then to use the map to guide a curtailed sampling programme with reduced laboratory testing. However, these economies depend on the quality of the classification. If the terrain under investigation can be resolved into clearly different land systems and land facets, recognizable both to the originator and to his successors, then the originator can label items of soil information by their facet and land system, and his successors can recognize other areas to which they are applicable. If the terrain does not lend itself to simple classification—and glaciated shields, some till landscapes and some areas of emergent coastline do not—then other means must be found for matching old to new sites.

BASIC MECHANISMS OF A DATA BANK

Three basic processes are involved in the operation of any data bank: input, storage, and retrieval. Some systems also provide for the manipulation of data in the store to generate new derivative information. The format of the input is governed by the type of storage and the main uses of the system, and in most systems field or remote sensing data must first be edited or processed into acceptable format.

The sorts of question which may be asked of geotechnical data banks are:

1. What soil profile may be expected at a particular point of interest?
2. What is the range of properties that may be anticipated of a particular soil profile or soil material in a particular area?
3. What sorts of engineering solution have been successfully applied to soils of a particular type?
4. Are there any problems or hazards involved in the application of engineering solutions or in the use of particular construction materials?
5. Where can a material of a specified type be found?

All of these questions contain the specifications for their answers—What point? What soil? What purpose? So every item of information put into the data bank must be labelled with the specifications of the engineering questions to which it may be the answer. Thus data must be entered, or *input*, in such a way that (i) the positions of observation points are indexed, (ii) terrain units with like soil profiles are indexed similarly, and (iii) soil properties, engineering solutions, and post-construction performance data are linked both to observation sites and to terrain units. The indexes provided by (i) and (ii) permit access to the data in (iii). Where the data bank is based directly on a terrain classification (e.g. on land systems and land facets, or similar units) the units of the classification can be used directly as 'pigeon-holes' or indexing characters. Each pigeon-hole accumulates stored information from sites on its terrain class, as a basis for generalizations about it and predictions about new sites on it. Once a terrain unit has been defined, an observer familiar with the classification system can identify the same unit elsewhere, either in the field or on aerial photographs; he does not need to refer to any map as his recognition is based on the matching of terrain types rather than on a map of their distributions. However, where the terrain does not lend itself to classification, because of obscuration by urban development or because of its inherent and unpredictable variability, meaningful spatial groupings can only be achieved if the terrain classes are recorded on maps, and data indexed on location. In this case the matching of new to old information is cruder but not impossible.

Data input forms have been specially devised for these purposes (e.g. Hodgson 1978). The forms themselves can be filed manually in boxes or folders, or the data recorded on them can be filed on cards or computerized. Computerized storage systems can easily combine or sort all data for a particular soil profile type, or terrain unit, on demand, or manipulate the data in other ways to produce a new data file. Unit boundaries on maps of terrain classes or base maps can also be digitized for computer storage. If data are indexed according to both terrain classification and location, the search for stored information about a new site proceeds as follows:

  (i) from a topographic map ascertain the co-ordinates of the site of interest;
 (ii) search the digitized terrain classification map to ascertain the terrain class at these co-ordinates—check this identification by any possible means;
(iii) search the stored data for information about the class now known to apply to the site of interest.

The output from a storage system may vary from direct retrieval of any raw data relevant to a request, to statistical summaries and map printouts where computer processing is used, according to its purpose. Webster (1977) gives examples of computer output.

EXAMPLES OF DIFFERENT DATA BANKING SYSTEMS

There exists a wide variety of geotechnical data banking systems but almost all of them store data in terms of either geological units or terrain units.

In urban areas, geological units provide the most appropriate basis for storage, while terrain units are better suited to regional data storage.

Geotechnical data banking for urban areas

In established cities and towns, where surface landforms have been largely obscured and modified by physical development, the only source of new soils and geological data is from site investigation records and from soil profiles exposed in the course of tunnelling, basement excavation, and foundation construction. These data may be used for the compilation of provisional geological and soil maps. The maps will continually be modified as more data become available. The map units provide the indexing or pigeon-hole units for geotechnical data. This basic approach is currently applied either formally or informally in London, Paris, Zurich, Moscow, Prague, Boston, Edmonton, Johannesburg, Tokyo, and Edinburgh (Legget 1973). The emphasis and approach vary according to local conditions; for example, mapping the network of old galleries resulting from the mining of gypsum is of prime importance in Paris.

De Beer and Biggs (1978) have described an operating data banking system for urban areas. It comprises a bedrock base map, an overlay map showing the sequence of soil horizons above bedrock and a summary data sheet, which provides the key to a store of micro-filed soil profiles and to the source of any quantitative geotechnical data on soil horizons similar to those present at the site of interest. Quantitative geotechnical data and information on foundation design are not stored directly within the system because such information is sometimes confidential and clearance must be sought from the organization concerned. The bedrock map (Fig. 12.1(a)) shows the distribution of various geological formations and structural features (faults, dips, etc.) and should include unconsolidated materials that have been recognized as stratigraphic units (e.g. London Clay). The overlay of soil horizons (Fig. 12.1(b)) shows their sequence, in terms of origin, without specifying their individual thicknesses or characteristics, which can be obtained from the micro-filed data. The locations of the original observation points are shown either on this overlay or on the bedrock map; these locations and their reference numbers provide a means of access to the micro-filed data and thence to the source of the quantitative geotechnical records. The summary data sheets (Table 12.1) list every record available in the bank in terms of a standard table of features. This provides a quick inventory of available data and may be computerized to facilitate the selection and combination of data relevant to particular sites, conditions, or problems.

An enquiry such as, 'What type of foundations are likely to be required for an office block with moderate column loads of about 50 tonnes each on site Z in Fig. 12.1(b)?', would involve the following search procedure:

1. The site is identified on the map and found to occupy the unit underlain by the soil profile PQ/RV/V (i.e. pediment deposits derived from quartzite, overlying residual lava grading into lava bedrock).
2. Fig. 12.1(b) also shows that the nearest site on the same soil profile for which a summary data sheet is available is site 310.
3. The summary data sheet for site 310 (Table 12.1) gives access to a site plan showing the locations of trial holes and to detailed descriptions of their soil profiles; it also records that the site investigation was carried out by Messrs Brown and Partners who performed triaxial shear and consolidation tests on undisturbed samples taken from the pediment deposit and from the residual lava.
4. The engineer now requests these results from Messrs Brown and Partners and ascertains that, while the sandy pediment deposit has a collapsible fabric, the underlying clayey residual lava can possibly provide adequate founding at a depth of about 8 to 10 m for displacement piles bearing the required loads.

Armed with this information the engineer is now able to plan a site investigation using the appropriate equipment, trial hole and sampling intervals, and a laboratory or *in situ* testing programme. Note that his original enquiry as to choice of foundation type, though typical of the sort of enquiry commonly put to the data bank, was not directly answered. The information provided was gleaned from a different, though similar, site and could therefore only serve as a guide to investigation procedures. This guiding function is, indeed, the main purpose of an urban geotechnical data bank; the need for individual investigation of every site can never be eliminated.

Geotechnical data banking based on physiographic terrain units

Brink *et al.* (1968) have described a terrain data bank which uses land facets (comparable to the terrain units of the PUCE system) as the basic units for classifying and indexing site records. These units can be readily identified either on aerial photographs of appropriate scale or in the field, and they are sufficiently small and homogeneous to be useful at project level, i.e. the disuniformity within a facet may not matter (Table 8.1, p. 188). The units of the system are used as pigeon-holes for storage, which is equivalent to making the list of class names serve as an index to the store. Data from any pigeon-hole can be applied to any site on a terrain unit that has been shown to be analogous, *either* from a terrain map, *or* through the interpretation of remote sensing imagery, *or* by field recognition. If the entry or key to a data bank is a terrain

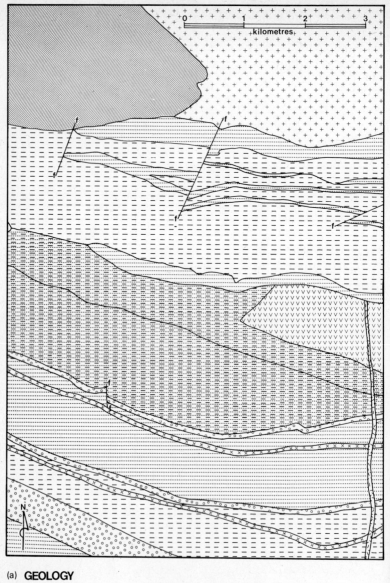

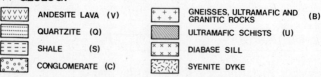

(a) **GEOLOGY**

ANDESITE LAVA (V)		GNEISSES, ULTRAMAFIC AND GRANITIC ROCKS (B)	
QUARTZITE (Q)		ULTRAMAFIC SCHISTS (U)	
SHALE (S)		DIABASE SILL	
CONGLOMERATE (C)		SYENITE DYKE	

FIG. 12.1. The bedrock map (a) assists in identifying the likely problems associated with residual soils in an urban area and, in conjunction with the overlay of soil horizons (b), affords access to the stored data by matching the site of interest to other sites of similar soil profile for which information is available.

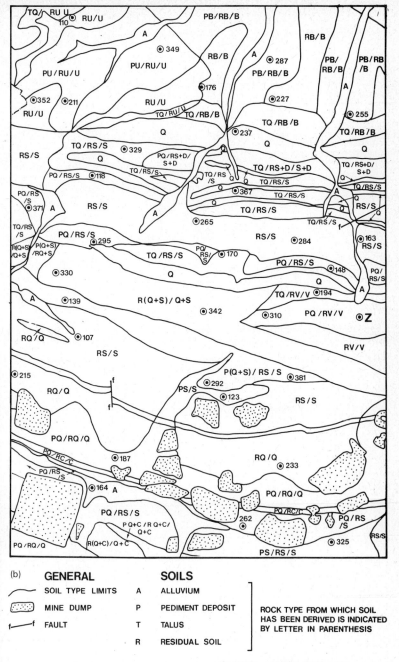

(b)

GENERAL		SOILS	
⌒ SOIL TYPE LIMITS	A	ALLUVIUM	
MINE DUMP	P	PEDIMENT DEPOSIT	
f———f FAULT	T	TALUS	
	R	RESIDUAL SOIL	

ROCK TYPE FROM WHICH SOIL
HAS BEEN DERIVED IS INDICATED
BY LETTER IN PARENTHESIS

PV/RV/V – Represents the profile :–
Pediment deposit derived from lava overlying
residual lava with bedrock underneath it.

TABLE 12.1

The summary data sheet for an individual urban site records the main features of the soil profile and the organization responsible for the investigation; it thus leads the enquirer who seeks information on a site of similar soil profile to a source of more detailed data.

(*After De Beer and Biggs 1978*)

GEOTECHNICAL DATA BANKING
INDEXING DATA SHEET

Made by: **N.H**   Date **1976/12/07**

Use only one character per space
For code information refer to Users Manual
On Card 1 Left justify all fields except 67–70   On Card 2 Right justify all fields

CARD 1

SUPPLIER & CALL NUMBER
1   3   4   11

`B&P` `I` `I`

SUPPLIERS JOB TITLE
12   29   CITY,TOWN
                                        30   32

`BIOSCIENCEBUILDING` `JNB`

MAP REFERENCE   GRID REFER   SUBURB REFER   STAND NO. PORTION NO.   CO ORDINATE OR 1ST STAND NO.   CO ORDINATE OR LAST STAND NO.
33   38   39   40   41   43   44   49   50   55   56   61

`1 8`   `B4`   `906`   `275`   `3 I 0`

DATE OF ORIGIN   AVAILABILITY CODE   TYPE OF STRUCTURE   TYPE OF DATA
62   65   66   67   70   71   73   75   77   80

`1976`   `5`   `07`   `TH`

CARD 2

TRANSPORTED SOIL WITH AVERAGE THICKNESS IN METRES
1   4

`05` `01`

PEDOGENIC MATERIAL WITH AVERAGE THICKNESS IN METRES
5   8

FIRST ROCK TYPE WITH DEPTH BELOW
GROUND LEVEL TO ROCK CONSISTENCY
9   12

`38`

FIRST STRATIGRAPHIC HORIZON
13

`47`

SECOND ROCK TYPE
15   18

`44`

SECOND STRATIGRAPHIC HORIZON
19

`47`

THIRD ROCK TYPE
21   24

`05`

THIRD STRATIGRAPHIC HORIZON
25

`35`

SHALLOW FOUNDATIONS
27

DEEP FOUNDATIONS WITH AVERAGE DEPTH IN METRES
29   32

MISCELLANEOUS KEYWORDS
33   52

`PEDIMENTDEPOSIT.LAVA`

53   72   73   75   76   80
                                      SUPPLIER & CALL NUMBER

`B&P` `00011`

classification then the terrain class of any new site must be identified before any information can be retrieved for it. Sometimes this is easy, sometimes it requires expertise. Usually the surveyor identifies the land system first, from the general environmental characteristics of the project area and airphoto interpretation. Then he identifies the land facet at any site of interest. So there must be a part of the system that provides the necessary information to assist the identification. This part consists of a number of recognition aids comprising:

1. A card index which records the bedrock geology, climatic zone, erosion surface, local relief, and physiography of every land system, usually according to a code. For example, the land system described in Fig. 12.2 might be indexed on the card as 'Basement schist and gneiss 463'. The attributes recorded on such an index card in no way define the land system, but merely give the surveyor quick access to a short list of possibly relevant or related land systems.
2. Representative annotated stereopairs of airphotos for each land system showing how land facets recur within it (Plate 1). These are sometimes supplemented by a table of recognition features (Plate 11). The relevant land system within the short list selected in (1) above is then identified on its pattern of land facets as revealed by airphoto comparison.
3. An idealized block diagram depicting facet inter-relationships (Fig. 10.4, p. 216).
4. A facet-index which summarizes the dominant characteristics of each land facet, including the sequence of horizons in the soil profile (Table 10.1, p. 218).
5. Any available ground photographs showing the surface appearance of the land facets.
6. It is also desirable to have a key map showing the distribution of land systems within any particular country or province. This map should be updated as land systems are defined and their boundaries delineated. The locations of any existing facet maps (or soil engineering maps) may also conveniently be indicated on such a key map. Where a land system map has been prepared and a site of interest can be located upon it the operator can gain access directly to its terrain class and hence to all relevant stored data; it must, however, be emphasized that a data bank based on a terrain classification system can operate effectively without such a key map.

The soil data which is stored includes detailed soil profiles for each record site. Data collected for individual soil horizons are entered on forms like Table 12.2. It is important that a locality sketch be included on a data storage sheet so that the site may be correctly identified when necessary. Where they are available,

FORM	LAND SYSTEMCARD		Security	U/C		ITEM No.
LOCATION	L.S. AREA	UGANDA	1 : 1M MAP	NA. 36		2063
	Lat. 0 ° ' "		Long. 32 ° ' " E			
LAND UNITS	LS	UG 72	LF			
	Name	MASAKA	LFLS			
CONTENT DESCRIPTORS	SOILS	GEOLOGY				
	LANDCOVER	APPEARANCE				
	RELIEF					SYSTEMCARD
Source	C.D. OLLIER	Org.	OXFORD		Date	15.11.65.

Climate : 1000–1300 mm. rainfall, bimodal; mild dry season.

Rock : Pre-Cambrian basement complex, mainly schists and gneisses mainly deeply weathered and lateritised.

Morphogenesis : Dissected old land surface in which massive laterite is preserved as level caps to major interfluves. Below these are long hill slopes leading to wide aggraded and frequently swampy valleys.

Soils : A variety of red loam lateritic (ferrallitic) type. (Buganda catena Kifu and Kaku series).

Vegetation : Forest/savanna mosaic with forest dominant along valleys.

Altitude : 1300 m. approx.

Relief : 120–150 m.

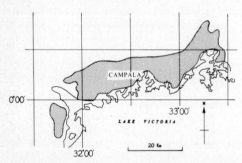

Stippled area shows extent of Masaka Land System.

Air Photographs : 15 : UGI2 : JUNE 1955 – 152.54 mm. : 24000' : 6" 012 and 013

FIG. 12.2. A land system class is indexed on environmental data to assist the separation of the few land systems, out of many, that could occur in a new area. (From MVEE).

TABLE 12.2

Data storage sheet for the geotechnical properties of soil materials occurring at one site within a land facet. These data are accessed through the classification system, i.e. once the facet at the site of interest is identified, data from all sites on other occurrences of that facet become applicable

A. INDEXING DATA

	Name	No.
Land system	KYALAMI	10
Land facet	CREST	1
Facet variant	EROSION SURFACE	1

	No.	Genesis and nature of material
Soil horizon	3	RESIDUAL GRANITE
Depth of sample	3.0 m	
Type of sample	DISTURBED	
Profile No.	1004	

	Job	Strip	No(s)
Airphoto	438	11	2686
Map reference	TC 250: 2628/2629		
Grid reference	2628 A K14		

Testing laboratory	TPA
Sample No.	L848/66 NIRR 211
Project	NIRR 9431-4232
Road & mileage/site	See locality sketch
Date	JANUARY 1967

TABLE 12.2 (*cont.*)

B.   TEST DATA

Parameter	Value
Liquid limit	37
Plasticity index	19
Linear shrinkage	9.3
%—1.5″	100
%—0.742″	100
%—0.078″	86
%—0.164″	60
%—0.029″	50
%—5 µm	
%—2 µm	
Grading modulus	1.04
Sand equivalent	
Activity (Skempton)	
SG of particles	
Mod. AASHO density (lb/Cu. Ft.)	121.8
Mod. AASHO OMC %	10.6
Standard AASHO density (lb/Cu. Ft.)	
Standard AASHO OMC %	
CBR at Mod. AASHO	38
CBR at Standard AASHO	
CBR swell %	1.38
Aggregate crushing value	
10% FACT	
Subgrade modulus (field)	
Compressibility and consolidation data (undisturbed)	
Coefficient of permeability { field	
compacted	
Others: (e.g. soundness tests, tests on stabilized materials, pH, clay minerals, soluble salts, percentage secondary minerals, particle angularity, Young's modulus, shear strength parameters)	

TABLE 12.2 (*cont.*)

C. CLASSIFICATION AND RATING

System	Standard value	Numerical index
Casagrande	CL	10
PRA	A-6(6)	44
Harvin	654	15
Others:		

LOCALITY SKETCH

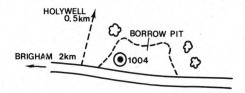

records of engineering solutions which have been applied in each facet, and their performance, should also be stored.

From time to time, as more input becomes available, the laboratory test data for each soil horizon in a land facet are statistically summarized in tabular form so as to provide information on average conditions and likely ranges of variation (Table 8.1, p. 188). This is best achieved by computer.

Questions 1 to 4 listed on p. 242 are readily answered by such a storage system and its index. If the system is also to answer question 5, i.e. where is a material of a specified type to be found, stored data on materials must also be indexed. Interim lists can be prepared which list all significant materials (e.g. good construction materials and problematical subgrade or foundation materials) and the land facets and land systems in which each is present.

When specific information about a distant area of interest is lacking, this method of terrain data banking can sometimes be used to make long range predictions. Such predictions are always a little hazardous (Mitchell, Webster, Beckett, and Clifford 1979), but subjective matches of distant terrain to a local analogue on geology, climate, erosion surface, relief, and physiography may be useful. The identification is best checked by comparison of airphoto patterns.

In the present form of this system, data are stored manually in a library of files each containing the data sheets for a single land facet. The files are grouped together in boxes each of which contains information on one land system. As the information in storage is extended, it may be necessary to house the files for each land system in a separate cabinet or even to resort to computer storage. The flow diagram in Fig. 12.3 shows the search procedure followed for any query.

CROSS-REFERENCING OF DATA

If it is desired to cross-reference different soil properties to other terrain data such as slope, altitude, aspect, geology, vegetation, and groundwater hydrology, then a much more comprehensive indexing system will be necessary. Extensive cross-referencing between different kinds of information demands either an exhaustive feature card index (Beckett 1971; Lawrance 1975) or computer manipulation. Such systems allow access to raw data, summaries or statistically manipulated material, either through the terrain classification itself, as described above, or via particular indexing characters such as location, soil type, soil property or engineering solution for which information has been stored and indexed. This flexibility is useful but is more complex and costly to achieve.

LIMITATIONS AND ADVANTAGES

The two working systems described in this chapter started small and so were designed for maximum ease of operation without a computer. They lack sophisticated cross-referencing and do not lend themselves to the retrieval of contextual data. For example, it would be a fairly laborious process to extract from the physiographic terrain data store all data that met the specification, 'information on the occurrence of ferricrete suitable for use as road base in areas of igneous rocks above an elevation of 2000 m', because the stored data is not specifically indexed for these features and, to answer the query, the search would have to proceed through each land system and land facet to find the relevant data. So far,

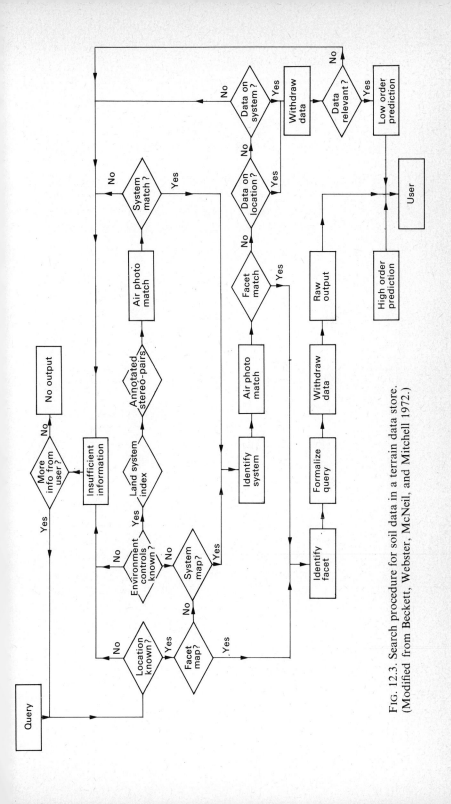

FIG. 12.3. Search procedure for soil data in a terrain data store. (Modified from Beckett, Webster, McNeil, and Mitchell 1972.)

contextual enquiries of this type have been infrequent; questions are about sites and not about abstractions. It is felt that the limitations are more than outweighed by the ease with which data can be stored and retrieved; if the mechanisms of any system are more complex than necessary the system tends to be neglected or abandoned.

A further limitation to the physiographic approach is its inherent degree of generalization: the units are, by design, only as uniform as the classes of terrain that can be distinguished on aerial photographs or in the field by the normal visual means available to the operator.

A limitation of the urban data storage system is its reliance on stratigraphy to extrapolate soil data. This disadvantage is somewhat outweighed by the fact that dense networks of point data are constantly generated in any urban area as development proceeds.

13. Afterword

It is a capital mistake to theorise before one has data.—Sir Arthur
Conan Doyle

At first sight a soil survey may appear to present a dilemma. On the
one hand, to be completely certain of the soil behaviour at every
site would require a close network of profile descriptions and tests
all over the project area. The price for complete certainty is, in fact,
infinite, since with soil nothing is ever wholly certain. On the other
hand the client wants to keep costs down to a reasonable level, but
without introducing unreasonable risks.

The engineering consultant must face up to this dilemma and
must not hide from the bugbear of litigation behind a screen of
expensive and unnecessary investigations. He must be his own
arbiter and choose a balance between reasonable cost and reasonable safety. His professional status requires him to be prepared to
make a balanced decision.

The art and skill of engineering soil survey lie in acquiring
sufficient information to ensure safe design but no more than is
really necessary. In a major project the surveyor will proceed by
successive stages:

(i) Explore the whole of the possible area rapidly in order to reject all but
a few possible sites or routes; if the location has been fixed by other
considerations this will not be necessary.
(ii) Investigate the most favourable locations to discover what particular
problems are likely to arise; this reduces the choice further.
(iii) Examine the selected location in *sufficient* detail (Fig. 13.1) to make
sure that the design is adapted to the whole range of conditions on the
site and that there will be no surprises for the contractor.

Chapters 7–12 of this book have described procedures of progressively smaller survey cost per unit area to serve projects of
progressively smaller hazard. The engineer must decide where on
this range his approach to a particular project must lie.

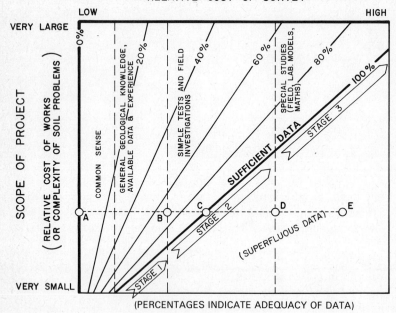

FIG. 13.1. The relative contributions of various aids in relation to the magnitude and complexity of a project. For a project of modest cost or complexity represented by A, just common sense and the application of existing knowledge would bring the engineer to B and provide him with 70 per cent of the data he requires for design. Efficient field investigations coupled with simple tests could yield adequate data (point C) for final design. Extending the field investigation unnecessarily to D would increase the cost of the survey without providing useful additional information. Undertaking special studies or tests, to E, would be totally unwarranted. On the other hand, special studies might legitimately comprise 40 per cent of the total cost of a very large and very complex project.

Appendix A.   Selected laboratory tests

This appendix outlines some of the more important tests for identifying or classifying soils for engineering purposes and for assessing their mechanical properties. The standards organizations of various countries (e.g. British Standards Institution 1975; American Society for Testing Materials 1974) have specified the detailed procedures.

TESTS TO DETERMINE PROPERTIES OF THE DISTURBED SOIL AND TO CLASSIFY SOILS

All the following tests may be performed on disturbed soil samples (Table A.1).

Indicator tests

(a) *Grading analysis*
Mechanical analysis, or the particle size distribution or grading analysis, of soils is carried out with a set of standard sieves, a sedimentation cylinder, and a hydrometer. With granular soils the sample is passed through a set of graded sieves, each of standard mesh size, and the fraction retained on each sieve is weighed. The finest sieve used in practice has a mesh opening of 0.074 mm, which is slightly larger than the limiting particle diameter between fine sand and coarse silt. Finer fractions are analysed by allowing the particles to settle through water. The rates of settling of different sizes are calculated from Stokes's Law, and their proportions are measured by regular hydrometer readings of the density of the suspension of soil particles in water at standard depth after appropriate times.

About 600 g of disturbed soil is adequate for sieve analyses, and about 100 g of the fine fraction for hydrometer determinations. The sample is prepared by standard methods which separate particles without breaking them down further, and disperse the aggregates of finer particles.

The results of the grading analysis are best presented in the form of a graph of the percentage by mass that is finer than any particle size against log (particle size), as in Fig. 2.1 (p. 8).

TABLE A.1
Some laboratory and in situ *tests to supplement a comprehensive soil profile description and the results of indicator tests, in order to assess the severity of a particular problem*

Property	Type of test	Sample type† and size (masses in kg)	*In situ* tests
	Laboratory tests		
Mineralogy	Optical microscope study	D 0.1	—
	X-ray diffraction		
	Differential thermal analysis		
	Electron microscope scan		
Particle specific gravity	Density bottle	D 0.05	—
Moisture content	Mass difference after drying	D, U, or C 0.3	Neutron moisture meter
Density	Mass per unit volume	U or C 0.3	Sand replacement, hydrodensimeter
Activity	Indicator, mineralogy, moisture content	D 1.0	—
	Density, double oedometer	U 10.0	—
Void ratio	Calculated from particle SG and density	—	—
Degree of saturation	Calculated from density and void ratio	—	—
Permeability	Permeameter	U or C 5.0	Piezometer, open borehole, well pumping, ring infiltrometer
Elastic modulus, shear modulus, Poisson's ratio	Triaxial	U 5.0	Seismic, pressure-meter, plate-bearing

TABLE A.1 (*cont.*)

Some laboratory and in situ *tests to supplement a comprehensive soil profile description and the results of indicator tests, in order to assess the severity of a particular problem*

Modulus of subgrade reaction	—	—	Plate-bearing
Compressibility	Consolidometer	U 5.0	Plate-bearing, pile loading Dutch probe, SPT, DCP
Sensitivity	Unconfined compressive strength	U 1.0	Swedish vane
Collapsible fabric	Density, consolido- meter with saturation under load, double oedometer	U 10.0	—
Shear strength	Triaxial, shear box	U or C 10.0	Swedish vane, pressuremeter plate-bearing
Ultimate bearing capacity	—	—	Dutch probe, SPT, DCP, plate-bearing, pile-loading
Unconfined compres- sive strength	UCS	U 5.0	—
Max. compacted density and optimum moisture content	Proctor or Modified AASHO compaction test	D 50.0	—
California bearing ratio	CBR	D 50.0	—
Soluble sulphates and chlorides	Conductivity, pH	D 0.3	—
Dispersion	Crumb test, % exchangeable Na, SCS double hydro- meter, pin-hole	D 10.0	—

† U: undisturbed (either natural or re-compacted); D: disturbed; C: compacted (either recovered 'undisturbed' from a compacted fill, or compacted to an equivalent density in a laboratory mould)

(b) *Atterberg limits*

Atterberg (1913) devised a number of simple physical tests to classify soils for agricultural purposes, but they have been more widely applied by engineers. The *liquid limit* (w_L or LL) is the moisture content at which the soil will just begin to flow when jarred slightly in the cupped palm of the hand; in this state it has a very low but measurable shear strength (about 1.7 kPa). At moisture contents between the liquid limit and the plastic limit a cohesive soil is plastic and will deform without cracking or crumbling. The *plastic limit* (PL or w_p) defines the boundary between the plastic and the solid state, and is the moisture content at which the soil begins to crumble when rolled into thin threads (about 3 mm in diameter) between the palm of the hand and a flat surface. In this condition it has a shear strength of about 170 kPa. The *plasticity index* (PI or I_p) is the difference between the liquid limit and the plastic limit and indicates the range of moisture content over which the soil remains in a plastic condition.

To determine the liquid limit and the plastic limit, about 300 g of the *soil mortar* (i.e. that fraction finer than about 0.4 mm) is needed. Originally the liquid limit was determined by adding water to the soil mortar to form a thick paste which was placed in a standard hemispherical bowl. The liquid limit was defined as that moisture content at which dropping the bowl 25 times through 10 mm caused the soil to flow together to close a standard groove cut in its surface.

A more accurate method uses a standard laboratory cone penetrometer; this is lowered several times into a dish containing a paste of the soil fines and the depths of penetration are recorded. The liquid limit is that moisture content corresponding to a penetration of 20 mm.

(c) *The linear shrinkage (LS)*

This is the percentage decrease in the length of a bar of soil on being dried in an oven from the liquid limit. It is a good indicator of the activity of a soil. The *shrinkage limit* (w_S or SL) is the moisture content below which further drying of the soil does not cause a reduction in its volume.

Mineralogy

The susceptibility of particles to chemical decomposition, and the presence of active clay minerals or soluble salts, may affect the engineering properties of a soil, but unless it is anticipated that there will be problems from these, mineralogical analyses are not usually carried out in engineering projects.

It may be useful to examine coarser soils under the optical microscope. Particles may be examined loose or they may be cemented together with a clear adhesive and then cut into sections in much the same way as thin sections of rock are prepared for petrographic analysis (Fig. A.1(a)). The very fine fraction of the soil may be studied by X-ray diffraction techniques using a goniometer. Different crystalline components, including clay minerals, are identified on their diffraction patterns. Differential thermal analysis (or DTA) produces a pattern of exothermic or endothermic response on heating a sample, which may be correlated with its mineralogical composition. Temperatures as high as 1200 °C may be required. An electron microscope may reveal the shape of individual clay crystals and aid their mineralogical identification (Fig. A.1(b)). This technique is particularly useful for revealing amorphous, yet active, components in the clay fraction which cannot be identified by X-ray diffraction.

A small sample of about 100 g suffices for all of these tests. The

(a)   (b)

FIG. A.1. (a). Photomicrograph of a thin section of aeolian sand with a collapsible fabric. Note thin colloidal coatings around the rounded sand grains. (b). 'Domains' formed by aggregates of clay particles show clearly under a scanning electron microscope. (After Oberholster and Brandt 1975.)

equipment and the procedures are described in standard texts on mineralogy.

However, field identification of the origin of the soil is often useful in predicting its likely mineralogical composition, and this helps to select the appropriate type of analysis.

Specific gravity (G)

The specific gravity of the soil particles is required in calculations of the void ratio, but it may also indicate the presence of unusual minerals. A common method, using a density bottle or pycnometer, determines the mass of a volume of water equal to the volume of the soil particles. Specific gravity is calculated as the ratio of the mass of the solid particles to the mass of the water. Three replicate tests on 50 g sub-samples should be carried out on each soil sample. Care is necessary to remove all the air from the voids between soil particles. The value for most soils is about 2.65, which is that of quartz, but many clays have specific gravities nearer 2.70. Fly-ash from the disposal lagoon of a power station may have a value of only 1.70, while the specific gravity of a beach sand containing significant concentrations of heavy minerals such as ilmenite may exceed 3.5.

It is sometimes difficult to obtain a consistent result with water if the soil contains soluble salts; if so, paraffin or some other liquid of known specific gravity may be used.

Moisture content (w)

The moisture content of a soil is the percentage of the mass of water to the mass of dry soil. Samples collected for this purpose should be sealed in airtight containers, or coated with a continuous layer of paraffin wax (p. 293). About 300 g is sufficient.

The mass of the sample is determined before and after drying it in an oven at about 105 °C for 24 h. Values of more than 100 per cent are possible, for example in a montmorillonitic clay.

Values of moisture content may prove a useful indication of likely soil behaviour. Note that a 20 per cent moisture content in sand would indicate a wet condition, while the same value in a clay would indicate only a slightly moist condition and the probability of a stiff soil consistency.

Moisture content may also be measured in the field with a neutron moisture meter (p. 295).

TESTS TO DETERMINE PROPERTIES OF THE UNDISTURBED SOIL FABRIC

The undisturbed samples required for these tests are collected by hand or in sampling tubes as described in Appendix B; sometimes it may be necessary to collect 'undisturbed' samples of compacted soils in embankments or other engineered fills, or equivalent samples may be prepared by compacting soil in a mould in the laboratory.

Density

The density of undisturbed soil is the mass of a unit volume of it. It may be measured in several ways, e.g. the mass of soil occupying a container of known volume may be determined; during road construction the density of a layer in the road prism is often measured by weighing the volume of sand required to replace a known volume of compacted soil (p. 296).

A sample of about 300 g is generally sufficient for laboratory determinations. The *bulk density* γ is the density at natural moisture content. For this the sample must have been sealed in the field as for moisture content determinations. *Dry density* γ_d is the mass of unit volume of the moist material after it has been oven-dried at 105 °C, i.e.

$$\gamma_d = \frac{\gamma}{\left(1 + \dfrac{w}{100}\right)}.$$

Saturated density γ_{sat} is the mass per unit volume of the material when saturated, i.e. when no air is present in the voids. This is usually the condition below the water table, when saturated density is synonymous with bulk density.

Submerged density γ' is often used in calculating effective stress and is the difference between the saturated density of the soil and the density of water, i.e.

$$\gamma' = \gamma_{sat} - \gamma_w$$

where γ_w is the density of water.

Void ratio (e)

This is the ratio of the volume of voids (both air and water filled) to the volume of solids in a soil. It is usually calculated from the dry density of an undisturbed soil sample and the specific gravity of the soil particles:

$$e = \frac{G\gamma_w}{\gamma_d} - 1.$$

The specific volume v is the total volume of soil occupied by a unit volume of solids, i.e.

$$v = 1 + e.$$

Degree of saturation (S)

This is the percentage saturation of the voids in a soil:

$$S = \frac{wG}{e}$$

or

$$S = \frac{\gamma_d}{\gamma_w}\left\{\frac{1+e}{e}\right\}w.$$

Permeability

The permeability of a soil is expressed as k, the coefficient of permeability, in the D'Arcy equation:

$$v = k\frac{h}{l}$$

v is the velocity and h/l is the hydraulic gradient, where h is the loss in hydraulic head and l is the length over which h is measured.

Soil permeability can be measured in the laboratory in a permeameter in which an undisturbed sample of 2 to 3 kg is subjected to a specified head of water and the volume flowing through it in a given time is measured.

For permeable soils such as sands (permeability about 10^{-3} m s^{-1}) a constant head is maintained in the test; for more impermeable soils such as clays (permeability of 10^{-9} m s^{-1} or less), the falling head test may be preferable; this measures the time taken for the water level in a stand-pipe to fall a measured depth.

It may be difficult to collect an undisturbed sample of non-cohesive soil, and disturbance of the edges of the sample may considerably affect the reliability of a laboratory test. For many purposes, therefore, soil permeability is best determined in the field (p. 297).

The results of these tests are mainly influenced by the initial moisture content of the soil, and special care should be taken to ensure that the sample is fully saturated (i.e. no air in the voids to interfere with the flow of water) before testing.

TESTS TO DETERMINE THE BEHAVIOUR OF THE SOIL FABRIC UNDER STRESS

All the tests given so far in this chapter measure static properties of the soil fabric; the following laboratory tests measure the behaviour of undisturbed (or recompacted) soil samples when subjected to stress. The samples are collected manually or in sampling tubes as described in Appendix B (pp. 279–81), or compacted material may be simulated by compacting soil material in moulds in the laboratory.

Elastic modulus or Young's modulus (E)

The value of this parameter is seriously affected by even slight disturbance of the soil. It is therefore difficult to determine in either the laboratory or the field, but the slope of the stress–strain curve of a loading and unloading cycle in the triaxial compression test on an undisturbed soil sample gives an approximate value of E (Fig. A.2).

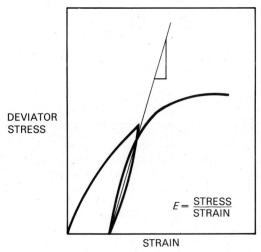

FIG. A.2. Young's modulus E is given by the slope of the stress–strain curve from a triaxial compression test.

However, the sample must be handled with great care to ensure that it is not over-stressed, and the test should be conducted in such a way that it will, as far as possible, re-impose the stress pattern which previously existed in the natural soil. This determination requires a cylindrical sample of the type used in the triaxial compression test described below.

The deformation properties of a soil, including the elastic modulus, can be determined *in situ* in the field by plate-loading tests or pressuremeter tests (p. 301), but their cost may not be justifiable except in the largest projects.

Poisson's ratio (v)

Values of this elastic constant vary from about 0.5 for incompressible materials such as saturated clay to about 0.15 or 0.3 for non-saturated soils. Only small displacements are involved and so Poisson's ratio is a difficult parameter to measure experimentally. It can sometimes be determined in a triaxal cell with a displacement transducer to record radial strain in the sample (Figs. A.3 and A.9). Cylindrical samples of the type used in triaxial shear tests are required, but once again the value determined is sensitive to disturbance produced by handling the specimen.

Compressibility

Compressibility is usually measured by means of a consolidometer (or oedometer) (Figs. A.4 and A.5), in which consolidation under various applied pressures is directly measured with a dial gauge or a displacement transducer. A triaxial cell is more convenient when the soils are stiff or stony. Water is expelled at a decreasing rate as load is applied to the soil and pressure is transferred to the soil skeleton; this increases the effective stress.

A small cylindrical specimen is needed for this test, which is usually carried out on a saturated sample. The analysis is divided into two stages; the first determines the relation between the void ratio and the applied pressure (Fig. 2.5(a), p. 18) and the second determines changes in consolidation with time (Fig. 2.4). A third set of procedures is used to study the swelling characteristics of the soil on removal of the load, and produces another graphical relationship between void ratio and applied pressure.

Although consolidation tests are usually carried out on saturated samples, it is sometimes useful to perform part of the loading cycle

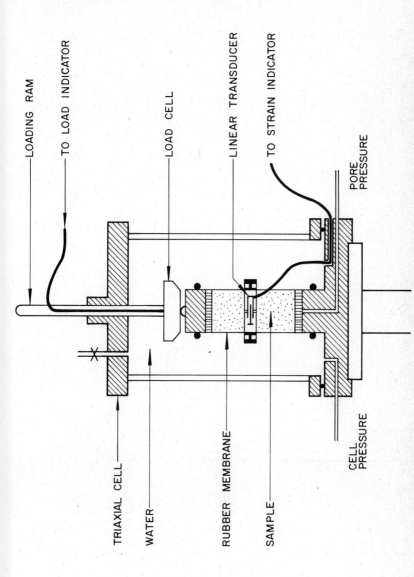

FIG. A.3. Triaxial test with a linear transducer to record radial strain (see Fig. A.9).

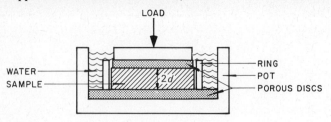

FIG. A.4. A consolidometer (or oedometer) measures volumetric changes in a soil under various loads or at different moisture contents.

FIG. A.5. Consolidometer.

on a sample in its natural moisture condition and then to saturate the sample at a pre-selected load. The direction and magnitude of the subsequent volume change reveal the susceptibility of the soil to collapse settlement or to heaving (pp. 118–21); rapid decrease in volume indicates a collapsible fabric, while a slow increase in volume indicates the presence of swelling clay minerals. All these tests are somewhat prolonged, since six or eight increments of load may be required to determine each relationship fully, and it may take twenty-four hours or more to reach equilibrium conditions after each increase.

All the relationships are plotted graphically (Figs. 2.4 and 2.5) and the following parameters can be obtained from the curves:

(i) *The coefficient of compressibility or volume change* m_v:

$$m_v = \frac{\Delta e}{\Delta p}\left(\frac{1}{1+e}\right)$$

where Δe is the change in the void ratio, Δp is the change in the effective pressure, and e is the initial void ratio.

(ii) *The coefficient of consolidation* c_v:

$$c_v = \frac{k}{m_v \gamma_w}$$

where k is the coefficient of permeability and γ_w is the density of water.

(iii) *The compression index:*

$$C_c = -\frac{e_1 - e_2}{\log_{10} p_1' - \log_{10} p_2'}$$

where e_1 and e_2 are the initial and final void ratios and p_1' and p_2' are the initial and final effective pressures.

Shear strength (τ)

The shear strength of a soil may be measured with a shear box or direct shear apparatus (Figs. A.6 and A.7). An undisturbed specimen is carefully trimmed to shape from a larger block or a tube sample before placing it in the apparatus. A normal load is imposed at right angles to the shear plane by directly loading the hanger, and the sample is then sheared along a pre-determined plane between the two halves of the box (Fig. A.6). The shear force is measured with a ring gauge or a load transducer; the horizontal displacement with time and the vertical displacement of the loading plate are also recorded. The rate at which strain is produced by the shear force is selected to achieve the soil condition required; a relatively rapid rate of about 0.5 mm min^{-1} may be used for an undrained test on clay, or a very slow rate of about 0.01 mm min^{-1} for a drained test. The residual shear strength of the sample is revealed by repeated reversals of the direction of loading and of the shear force after there has been a displacement of about 10 mm in the sample.

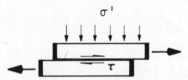

FIG. A.6. The shear strength of a soil may be determined in a shear box apparatus.

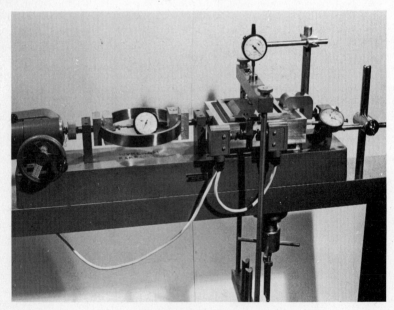

FIG. A.7. Shear box apparatus.

The specimen is usually kept inundated with water during the test. There is no means of controlling or measuring pore-water pressures during rapid tests, and it is assumed that pore-water pressures are dissipated completely during slow tests.

Shear strength may also be measured in a triaxial apparatus, where a cylindrical specimen is subjected to controlled radial pressure while the axial pressure is increased (Figs. A.8 and A.9).

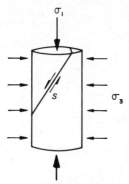

FIG. A.8. The triaxial test to determine the shear strength of a soil.

The resulting shear stress on any plane through the sample is calculated from the combination of these two stresses. The advantage of this apparatus over the shear box is that the pore pressure within the sample may be measured, and any pre-selected stress path may be followed by controlling the radial and axial pressures. The rate of strain can also be controlled, but reversals to measure residual strength are not possible, and residual strength can be determined only along a pre-existing or pre-formed failure plane cut in the sample.

The results of a shear strength test in a shear box or a triaxial cell are plotted graphically, and values of cohesion and angle of shear resistance are obtained directly from the graph (Fig. 2.7, p. 24).

Unconfined compressive strength (UCS)

The compressive strength of any material is the maximum stress it can withstand when compressed between two parallel plates. The use of the term 'unconfined' here emphasizes the distinction between this test and the triaxial test, in which a confining stress is applied at right angles to the major compressive stress. The test is

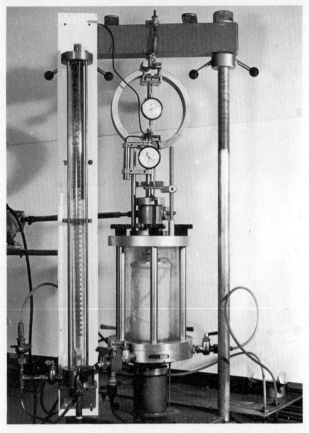

FIG. A.9. The triaxial apparatus may be used to determine the shear strength of a soil and, less reliably, to determine Young's modulus, the shear modulus, and Poisson's ratio.

carried out, on cylindrical samples with length equal to about twice the diameter, in a compression test apparatus, and the strain is applied sufficiently fast to cause failure in about ten minutes. The procedure is rather crude and any imperfections in the sample such as fissures or joints will initiate early failure at a low stress. The test is, however, useful on intact saturated clay, where the results can be related directly to consistency. Under ideal circumstances the cohesion of a soil is equivalent to half its unconfined compressive strength.

TESTS TO DETERMINE THE COMPACTION PROPERTIES OF SOIL

The tests described in the previous section are usually conducted on undisturbed samples, although soil material to be used in earth dams or other engineered fills can be compacted in moulds to simulate the compaction which will be achieved in the field. This section deals mainly with tests to measure the compaction characteristics of soil materials during the construction of earthworks. Relatively large disturbed samples of about 50 kg are necessary, since each determination requires about 10 kg and several determinations are usually required for each sample.

Moisture density test

A standard compactive effort is employed to simulate the effect of heavy field plant and to determine the relationship between the moisture content of the soil and the density achieved during compaction. One standard test, often referred to as the Proctor test or standard AASHO test or British normal effort,† provides for a metal hammer of 2.5 kg mass (5.5 lb) to fall 310 mm (12 in) onto soil in a mould of 100 mm (4 in) diameter and 116 mm (4.6 in) height. Three equal layers are given 25 blows each to impart $12\,500\ kN\ m^{-3}$ impact energy to the soil. This standard is often used to specify compaction in embankments (e.g. earth dams) where the soil must be compacted to moderate density in fairly thick layers (up to 300 mm).

Another standard known as the modified Proctor test or modified AASHO test† employs a similar but heavier hammer of 4.54 kg mass (10 lb) falling 416 mm (18 in) onto the soil in a mould of 150 mm (6 in) diameter, with five layers and 56 blows per layer. This imparts $56\,000\ kN\ m^{-3}$ impact energy to the soil in conformity with the compactive capacity of the heavy plant used in most contemporary earthworks projects. The British heavy effort test imparts an equivalent input of energy to soil in a 100 mm mould. This standard is used to specify compaction in the controlled layers of roads and runways or in the sub-foundation materials for buildings, where the soil must be compacted to high density in fairly thin layers (up to 150 mm) to support relatively high loads.

† See footnotes on p. 36.

The density achieved by compacting a given material depends upon its moisture content. A relatively dry soil resists compaction and a low density results; increased moisture causes lubrication and leads to higher density, and an excess of moisture prevents the particles from attaining a state of close packing. A moisture/density curve is plotted from a set of determinations; its peak value is the *maximum dry density* and the corresponding moisture content is the *optimum moisture content*. Both of these values depend on the compactive effort and also on the type of compactive plant used. The density achieved during field compaction is often expressed as, for example, 95% Mod. AASHO or 98% AASHO, meaning a dry density which is 95 per cent of the modified AASHO maximum dry density or 98 per cent of the standard AASHO maximum dry density.

California bearing ratio test

The California bearing ratio test is an empirical test which was devised some forty years ago to evaluate the strength of compacted road material. The results cannot be related to any fundamental soil strength parameters, but they provide a useful means of comparing different types of material. The test procedure must be strictly adhered to in order to relate the results to standard methods of road foundation design. Usually the sample is compacted at the

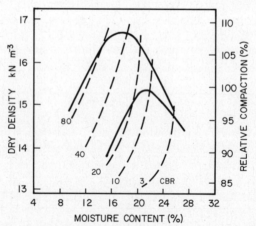

FIG. A.10. The CBR characteristics of a sandy clay superimposed on the compaction curves of Fig. 2.12 (p. 31).

desired effort (mostly between 90% and 98% Mod. AASHO) in a mould of 150 mm diameter as in the moisture–density test, and soaked in water for 4 days before a penetration test is carried out. A standard plunger of 19.35 cm^2 (3 in^2) cross-sectional area is advanced into the surface of the compacted soil at a rate of 1.25 mm min^{-1}. The load required to achieve 2.5 mm penetration, expressed as a percentage of a standard load of 150 kN (3000 lb force) is the CBR value of the compacted material.

The test is often carried out on sub-samples compacted to different densities at different moisture contents; Fig. A.10 shows a typical pattern of CBR strengths superimposed on compaction curves.

Appendix B. Soil sampling and *in situ* testing techniques

SAMPLING TECHNIQUES

The techniques by which the soil surveyor obtains samples to reconstruct the soil profile or for laboratory testing vary greatly in the extent to which they disturb the natural fabric of the soil or change its moisture content. Those which can recover samples in a relatively undisturbed condition are preferred for the most detailed site investigations (Chapter 7), but it may not be possible to take 'undisturbed' samples from some soils (e.g. cohesionless sands) by any method, nor is it always possible to justify the expense of taking undisturbed samples from every profile in an investigation, especially where a shallow water table prevents the investigator from descending the holes and taking the samples by hand. So the investigator often has to use alternative or cheaper sampling methods which disturb the soil fabric. Various sampling techniques are described below in general order of their increasing disturbance of the soil and decreasing cost:

Hand sampling

Manual sampling is possible in augered trial holes of large diameter or in test pits excavated by a mechanical excavator or by hand. In soils which are not fissured and have at least some cohesion, a block can be detached by cutting grooves around the sides of a 200–300 mm cube of soil with the sharp end of a geological pick, and then carefully prizing it away from the side of the hole. Where the soil is fissured and tends to fragment, a sharp-edged steel ring of 200–300 mm diameter is driven into the side or the bottom of the hole, and the surrounding soil is then cut away with a geological pick. In very stiff soils it is often impossible to insert the ring without the assistance of a hydraulic jack acting against a steel plate on the opposite face of the hole. Disturbance of the sample during its excavation and removal is usually restricted to the outermost 50 mm.

Sampling with thin-walled tubes

Thin-walled tubes are used to sample clayey soils in boreholes which are too small to be descended and sampled manually. A tube is said to be thin-walled when the 'area ratio' (ratio of the cross-sectional area of the tube walls to the cross-sectional area of the sample) is less than 10 per cent (Hvorslev 1949). A thin-walled sampler like the *Shelby tube* (Fig. B.1) is usually made of steel and is

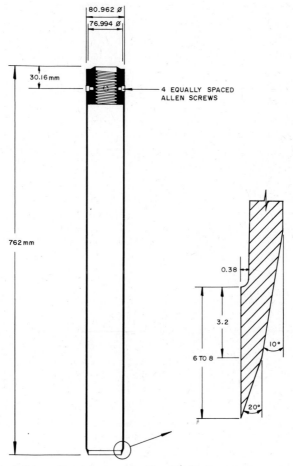

FIG. B.1. A thin-walled tube sampler like this Shelby tube is used to collect undisturbed soil samples.

equipped with a sharp cutting edge and a head which is threaded for attachment to the rods of a rotary drilling rig so that it can be lowered down a borehole. The hydraulic mechanism of the drilling rig is used to force the tube as quickly and as smoothly as possible into the soil below the bottom of the borehole, taking care not to compress the sample by pushing the tube more than its own length. The sample is detached by rotating the tube slowly and, after withdrawing the tube from the borehole, the sample should be extruded in the same direction as it was taken.

A thin-walled *stationary piston sampler* (Fig. B.2) is used to

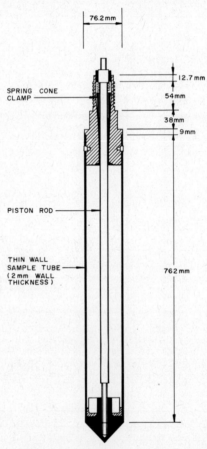

FIG. B.2. A thin-walled stationary piston sampler, suitable for very soft cohesive soils.

collect samples from very soft cohesive soils. The piston is attached to a separate rod from the surface so that it can be held in place as the tube is pushed into the soil to prevent soft soil from flowing into the tube and contaminating the sample. It also ensures retention of the sample during recovery.

Sometimes a soil profile is described from a continuous succession of tube samples. Where the soil is not sampled continuously the borehole is deepened in stages using one of the techniques described below and each new horizon for sampling is identified from the disturbed soil brought up from the bottom of the hole. Table B.1 demonstrates a form designed to summarize data on soil samples recovered in the course of an engineering investigation.

Provided that they are inserted quickly and smoothly and are not 'overdriven', thin-walled tubes disturb only the outermost few millimetres of the sample. However, if the borehole has been deepened by the rotary drilling or wash-boring techniques described below, the drilling fluid may have increased the moisture content of the underlying soils. In contrast, thick-walled samplers, such as the U-100 tube, in which the area ratio exceeds 10 per cent, produce considerable disturbance and the samples should not be used in assessments of the properties of the undisturbed soil fabric.

Other techniques for recovering undisturbed samples

Several devices can recover undisturbed samples from materials such as very stiff clays or pedocretes (p. 65), which cannot be sampled with thin-walled tubes without damaging the tube or severely distorting the sample. The *Denison sampler* (Fig. B.3) consists of a stationary inner tube which extends in front of a rotating outer tube, equipped with a cutting bit. The bit is cooled by circulating drilling fluid in the same way as the rotary core drill (see below). This configuration ensures that the sample recovered in the inner tube suffers the minimum of distortion or contamination by drilling fluid (Acker 1974). The entire assembly is attached to the rods of a core-drilling rig, which provides the rotary action.

Rotary core drilling

Rotary core drills (Fig. B.4) can provide continuous samples from deep boreholes, and are used when the soil is very resistant or when rock strata must be penetrated. A hollow, rotating tube at the end of the drill rod is equipped with a hard cutting edge, or bit, of

Table B.1
A soil sampling field sheet

FIRM GEOTECHNICAL SURVEYS

PROJECT TITLE Site investigation for Binis Bantu

NOTES: Thunderstorm with 10 mm rain last night.

SITE Old Market Place, Station Road, Waltown

WEATHER Fine

SAMPLER'S NAME: A.J. Garner

GROUND LEVEL 1438.65

WATER TABLE DEPTH 4.80 m 72/09/30

PAGE 3

DATE 72/09/28

BORE HOLE No. BR6

TIME	DEPTH AUGER FROM	TO	TAPED DEPTH HOLE	TUBE PUSHED FROM	TO	LENGTH OF DRIVE	NETT PUSH PRESSURE	TIME OF PUSH (MINS)	COMMENT ON RESIST	TAPED DEPTH NEW HOLE	SAMPLE LENGTH	WAXED LENGTH FROM	TO	SAMPLE NO.	DESCRIPTIONS AND GENERAL — VANE TESTS (SIZE / DEPTH / PUSH PRESSURE / PEAK / RESIDUAL) — REMARKS	PROFILE SYMBOLS
08 4.00	–	–	–	0.00	0.50	0.50	2.000	14 sec	Hard	0.50	–	–	–	89	Disturbed sample in bag.	
08 4.15	0.00	0.50	0.50	0.50	1.00	0.50	1.000	7	Stiff	1.00	0.41	0.59	1.00	25	Undisturbed sample in core.	
08 4.30	0.50	1.00	1.00	1.00	1.50	0.50	1.500	7	"	1.50	0.43	1.07	1.50	26		
08 4.45	1.00	1.50	1.50	1.50	2.00	0.50	1.400	6	"	2.00	0.48	1.52	2.00	27		
09 4.00	1.50	2.00	2.00	2.00	2.50	0.50	1.300	6	"	2.51	0.51	2.00	2.51	28		
09 4.15	2.00	2.51	2.51	2.51	3.00	0.49	1.300	5	"	3.00	0.49	2.51	3.00	29		
09 4.45	2.51	3.00	3.00												30x60 3.40 2.00 6.30 1.85 VANE	
12 4.00	–	–	–	3.00	3.50	0.50	2.400	15 sec	"	3.50	0.40	–	–		Block on vane test OK.	
12 4.30	–	–	3.50												30x60 3.90 3.00 6.90 2.65 VANE	
15 4.30	–	–	–	3.50	4.00	0.50	2.300	8	"	4.00	0.45	–	–		Block on vane test OK.	
16 4.00	4.00	4.55	4.55	4.55	5.05	0.50	1.500	7	"	5.05	0.49	4.56	5.05	30	Undisturbed sample	
16 4.30	4.55	5.05	5.05	5.05	5.50	0.45	1.900	7	"	5.50	0.45	5.05	5.50	31	"	
17 4.00	5.05	5.55	5.50	5.50	6.00	0.50	1.400	5	"	5.99	0.48	5.51	5.99	32	"	
17 4.30	5.55	6.00	6.00	6.00	6.50	0.50	800	5	"	6.50	0.50	6.00	6.50	33	"	
															END OF HOLE	

tungsten carbide or a 'crown' containing diamond chips, to penetrate resistant materials. A flushing fluid (usually water or drilling mud) cools the bit and removes particles dislodged by it. A sample, or core, of the material penetrated is retained in the hollow tube, or core-barrel; how far the sample is disturbed by the flushing fluid and the rotation of the core-barrel depends partly on the hardness and cohesion of the soil, but more especially on the design of the barrel. With a single-tube barrel (Fig. B.5) the flushing fluid flows over the surface of the sample and may wash particles from sandy or soft materials; the fabric of such materials is easily destroyed by contact with the rotating core-barrel. The sample is

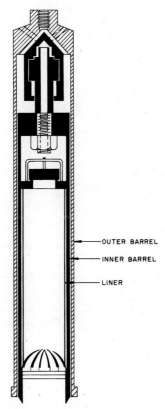

OUTER BARREL

INNER BARREL

LINER

FIG. B.3. Denison double tube core-barrel for obtaining undisturbed samples of stiff clays and cemented materials.

less disturbed in double- and triple-tube barrels (Fig. B.6), in which the inner tube remains stationary and separates the sample from the drilling fluid.

If the sides of the borehole tend to collapse a thick drilling fluid or mud, usually made from bentonite and water, is used to infiltrate the soil and helps to stabilize it. If this fails a steel tube, or casing, is used to line the hole. Such casing is also used for support in the

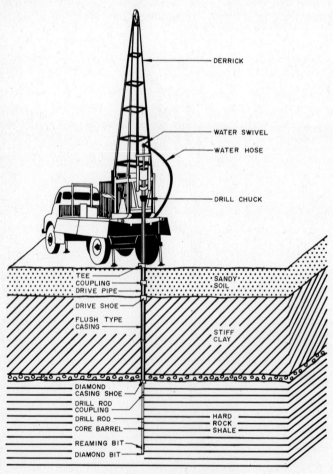

Fig. B.4. Rotary-core drilling rig.

shell-and-auger, rotary percussion drilling, and wash-boring techniques described below. Since the casing is designed to fit inside the initial borehole, core-barrels of smaller diameter are used to deepen the hole through any casing in the upper levels; alternatively, the hole is enlarged by 'reaming' so as to accept a wider casing. Table B.2 lists the more common sizes of core-barrels and casings used in rotary core drilling.

The principal disadvantages of soil sampling by rotary core drilling are its high cost and slow rate of progress, and the degree of disturbance suffered by the core, particularly when soft or loose soils are penetrated. This method is best suited to the recovery of rock samples.

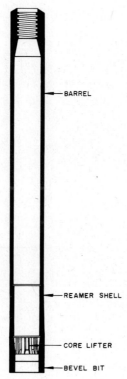

BARREL

REAMER SHELL

CORE LIFTER

BEVEL BIT

FIG. B.5. A single tube core-barrel is effective in rock sampling but severely disturbs the fabric of soils.

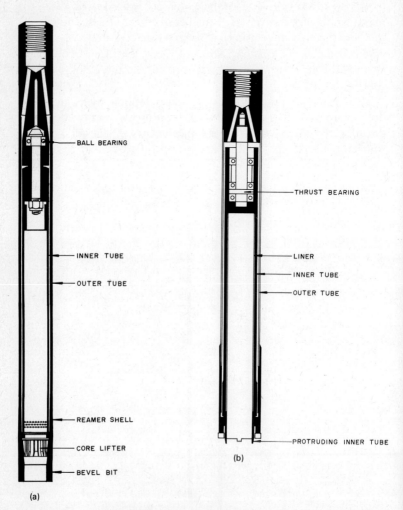

BALL BEARING

INNER TUBE

OUTER TUBE

REAMER SHELL

CORE LIFTER

BEVEL BIT

(a)

THRUST BEARING

LINER

INNER TUBE

OUTER TUBE

PROTRUDING INNER TUBE

(b)

FIG. B.6. The double tube core-barrel (a) and the triple tube core-barrel (b) cause somewhat less disturbance to the soil fabric than the single tube barrel.

TABLE B.2

Common core-barrel, core and casing sizes used in rotary core drilling

Size	OD‡ (mm)	Core diameter (mm)	Size	OD (mm)	Core diameter (mm)
†EWM	37.7	22.2	EWT	37.7	23.0
†AWM	48.0	30.1	AWT	48.0	32.6
†BWM	59.9	42.0	BWT	59.9	44.5
†NWM	75.7	54.7	NWT	75.7	58.7
			HWT	98.80	80.94

† Previously called NXM, BXM, etc.
‡ Outside diameter.

Size	OD (mm)	Core diameter (mm)	Type	OD (mm)	Core diameter (mm)
T2-46	46	32	Denison	74.6	47.6
T2-56	56	42	triple	88.9	59.9
T2-66	66	52	tube	101.6	71.4
T2-76	76	62		139.7	104.0
T2-86	86	72		196.9	160.3
T2-101	101	84			
T6-116	116	93			
T6-131	131	108			
T6-146	146	123			

Casing

Size	OD (mm)	ID (mm)	Fits the holes made by the following core barrels without 'reaming' to enlarge the diameter
EX	46.0	41.3	AWM, AWT, T2-56
AX	57.2	50.8	BWM, BWT, T2-66
BX	73.0	65.1	NWM, NWT, T2-76
NX	88.9	81.0	HWT, T2-101
HX	114.3	104.8	T6-116

Sampling with small-diameter augers

Small-diameter augers for deep sampling most frequently use a spiral bit of about 75–350 mm diameter (Fig. B.7), which is rotated by the kind of rig used for core drilling. Disturbed soil samples are recovered from the spiral blade as it is withdrawn from the borehole. The hole is deepened by adding new sections to the auger or to the rods. Because no drilling fluid is used, augering provides a useful means of sampling in areas with a deep water table and when a knowledge of the natural moisture content of the soil is critical. Some mechanically operated augers have hollow stems which allow undisturbed samples to be recovered in thin-walled tubes without withdrawing the blade. Since these tools function by the continuous penetration of the auger into the soil, the sides of the borehole cannot be supported by casing, and the technique is useful only in self-supporting, cohesive soils. *Shell-and-auger* boring is useful in non-cohesive soils: a steel casing or shell with an internal diameter of about 150 mm supports the sides of the borehole and accommodates a rotating hollow tube equipped with a cutting edge; behind the cutter a flap-valve gives access to the tube and prevents cohesionless materials such as sand or gravel from falling back into the borehole (Fig. B.8). The auger tube can be replaced by a variety of chopping bits to advance the borehole through bouldery or cemented soils.

Soil can also be sampled with a *hand-auger* (Fig. B.9). The Hardy pick auger is useful in sandy soil since it incorporates a bucket to retain the sample, whereas the Edelman auger offers rapid penetration of cohesive soils, which adhere to the spiral blades. Hand augers with a standard shaft can penetrate about 1.25 m, but this depth can sometimes be extended to 10 m or more by attaching threaded extension rods. A tripod and block-and-tackle must be used to lift the auger from the hole after each advance, or the weight of the extensions may prove unmanageable.

Spoon and channel samplers

The standard penetration test, from which the properties of the soil are assessed *in situ* (p. 166), measures the number of blows required to drive a *standard split-spoon sampler* or Raymond spoon into the soil (Fig. B.10). Other spoon samplers are available. As the soil is penetrated a disturbed sample is forced into the hollow spoon; a

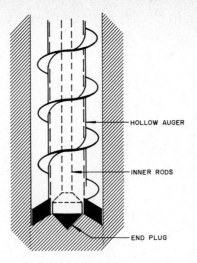

HOLLOW AUGER

INNER RODS

END PLUG

ADVANCING INTO GROUND BY ROTARY DRILLING

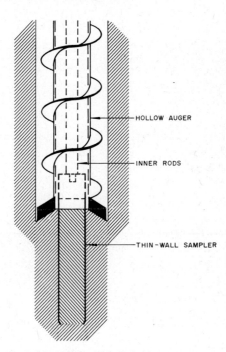

HOLLOW AUGER

INNER RODS

THIN-WALL SAMPLER

SAMPLING TUBE PUSHED INTO THE SOIL

FIG. B.7. The hollow stem auger permits samples to be collected in thin-walled tubes without withdrawing the auger spiral from the borehole.

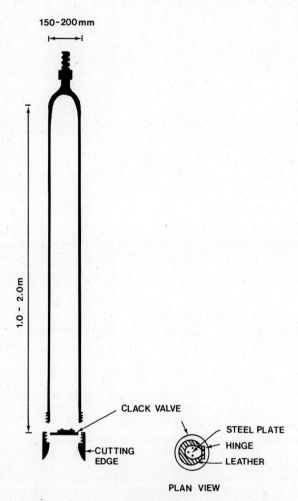

150-200 mm

1.0 - 2.0 m

CLACK VALVE

CUTTING EDGE

STEEL PLATE

HINGE

LEATHER

PLAN VIEW

FIG. B.8. With the shell-and-auger technique a sand baler is used to deepen a borehole in non-cohesive soils. The sides of the hole are supported with a hollow metal casing (or shell) through which the baler (or auger) is rotated.

sample may be retrieved between each penetration provided that the cohesion is not so low that the soil falls back into the hole when the sampler is withdrawn. The standard penetration test is usually carried out in cased boreholes, which are deepened between penetrations by augering or wash-boring.

Other devices for testing the *in situ* properties of soil, such as the Dutch probe or the dropweight cone penetrometer (pp. 314–23), in which a metal cone is driven into the soil, sometimes retrieve disturbed soil samples in open grooves or channels cut into the side of the cone. Such sampling devices are ineffective in cohesionless soils.

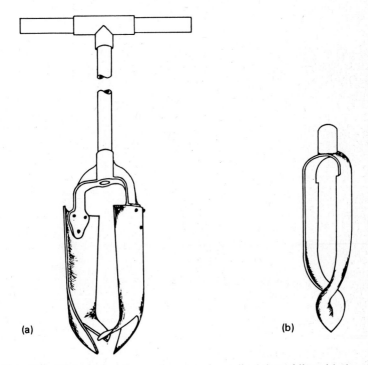

(a)

(b)

FIG. B.9. Disturbed soil samples may be collected rapidly with hand augers; the Hardy pick type (a) is useful in sandy soils while the Edelman (b) is better in clays.

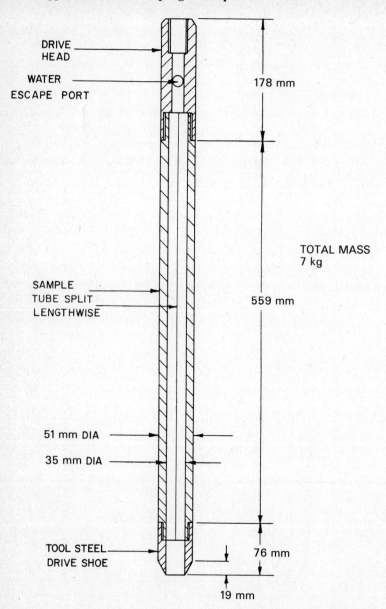

DRIVE
HEAD

WATER
ESCAPE PORT

178 mm

TOTAL MASS
7 kg

SAMPLE
TUBE SPLIT
LENGTHWISE

559 mm

51 mm DIA

35 mm DIA

TOOL STEEL
DRIVE SHOE

76 mm

19 mm

Fig. B.10. A standard split-spoon sampler, or Raymond spoon, collects disturbed soil samples during the execution of a standard penetration test.

Sampling with a rotary percussion drill

Rotary drills (Fig. B.11) use compressed air to rotate a hard metal bit, or hammer, while simultaneously impacting it rapidly against the bottom of the borehole. Casing can be used to support the sides of the hole as it is deepened. Since the fine fragments, or cuttings, produced in this way are flushed to the surface by compressed air, foam or water, the sample is not only highly disturbed, but may also be sorted so that the various particle sizes are not recovered in their true proportions.

Sampling by wash-boring

Wash-boring provides an inexpensive means of deepening a borehole by discharging water under pressure through nozzles in a chopping bit. The water jets loosen the soil and flush it to the surface, while at the same time a hollow casing is driven into the soil with a drop-weight to support the sides of the borehole. As with rotary percussion drilling, the sample which reaches the surface is often sorted, and the moisture content of the undisturbed soils at the bottom of the borehole will have been increased unless they were already saturated. This may affect the results of any tests carried out *in situ* or on undisturbed samples.

PRESERVATION AND STORAGE OF SAMPLES

Undisturbed samples must be transported and stored with care: shocks may disturb the soil fabric and large variations in temperature may alter the moisture content and the soil chemistry. Undisturbed samples recovered at natural moisture content must be sealed against moisture loss pending laboratory testing. Blocks cut from the sides of test pits and trial holes are coated immediately with about 15 mm of paraffin wax. Tube samples are slowly and evenly extruded in the same direction as they were taken, enclosed in plastic film, and then sealed into tubular containers with paraffin wax. Where possible, undisturbed samples should be kept in a specially constructed store-room at a high humidity to prevent any possible evaporation through imperfect or damaged wax coatings.

Disturbed samples are usually kept in plastic bags, but when their moisture content is to be measured they are sealed in airtight tins or plastic jars.

FIG. B.11. Highly fragmented and sorted particles may be retrieved using a rotary percussion drilling rig. Fragmentation and penetration of the soil or rock is achieved by repeated blows of a 'hammer', one type of which is shown. (Photographs by courtesy of Messrs. F. B. Drilling (Pty.) Limited and the Ingersoll Rand Company.)

All samples must be labelled both on the inside and the outside of their containers with a clear record of their provenance that cannot be damaged by moisture. It is sometimes useful to store samples in double plastic bags; the label is inserted between the inner and the outer bags to protect it from damage and wetting.

In situ TESTING TECHNIQUES

Samples taken from pits and from boreholes are seldom entirely undisturbed and may not be wholly representative of the soil horizons from which they were collected; and sampling may pose severe problems in non-cohesive sands or soft and sensitive clays. In theory there are great advantages in testing soils *in situ*. In practice many such tests give only indirect and approximate measures of soil properties.

The properties of greatest importance to the engineer have been discussed in Chapter 2, and Appendix A describes their measurement in the laboratory. Those which can be measured *in situ* in the field are considered below.

Moisture content

The moisture content of natural soils and fills can be measured to shallow depth from the soil surface with a *neutron moisture meter*. Several devices are marketed under different names. All contain a source (often 50 mCi americium–beryllium which has a half-life of 400 years) and a detector (such as a boron trifluoride tube): the source emits high energy neutrons into the soil and the detector measures the return of the low energy neutrons that have been scattered back at slower speed by collision with hydrogen nuclei in the soil water. The rate of return of these neutrons can be calibrated against the moisture content of the soil.

A similar radioactive technique with a source of gamma-radiation (such as 8 mCi caesium-137, which has a half-life of 33 years) and a Geiger counter can be used to assess the density of the soil. The amount of back-scattered gamma radiation can be calibrated to give the soil density. The two devices can be combined into one unit (Fig. B.12) with a source which emits both fast neutrons and gamma rays (such as a 10 mCi radium–beryllium source which has a half-life of 1620 years). In both devices the particles are emitted either from mechanical probes lowered into

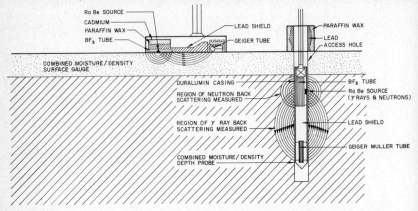

FIG. B.12. Neutron and gamma radiation may be used to measure the moisture content and density of *in situ* soils to shallow depth.

the soil down aluminium access tubes of small diameter or from a source placed on the flat soil surface. The latter configuration is useful for measurements on compacted fills or layers placed to form the road prism, where frequent determinations are needed to ensure that the compaction of the layers conforms to the specifications of the design. However, it is not usually possible to make measurements to depths greater than about 500 mm with the surface device, but a probe may be lowered to any depth to which the aluminium tube has been installed.

Some care is needed in the use of these radioactive devices, but contemporary models incorporate adequate shielding. Operators should always wear special monitoring film clips which are subject to regular checks, to meet health requirements.

Density

The density of the soil can be measured *in situ* with the hydro-densimeter just described, or it can be determined by the *sand replacement* method. As with the hydrodensimeter, this method is useful only to shallow depth, and is most widely applied to check the density of compacted soils.

In this method a column of soil less than 100 mm in diameter is excavated with chisels and scoops to a depth of about 100 mm. The mass of the soil removed is determined, together with its natural moisture content; the hole is then filled with clean, dry, single-sized, freely running medium sand from a special cylindrical container, and the loss in mass of the container is measured. The bulk density of the sand is known and is used to calculate the volume of the hole and hence the bulk density of the soil.

The volume of the hole can also be determined with an apparatus which measures the volume of air required to inflate a balloon to fill it.

Permeability

The coefficient of permeability of a soil can be determined in several different ways. A *hydraulic piezometer* can be used to introduce water into the soil and so to measure its permeability, and it can also be used for long-term measurements of pore-water pressure in the soil. To measure the permeability of a particular soil horizon a piezometer (Fig. B.13) is installed in a borehole drilled to the appropriate depth and lined with a perforated casing; a sand pocket 300 to 600 mm long is placed around its tip and the sand is capped with an impermeable disc. A grout consisting of a mixture of cement and bentonite is then pumped into the borehole above the disc to prevent water from rising in the hole and entering higher soil horizons. Once the piezometer is installed, water is passed through its tip at constant pressure, and the rate of outward flow through the sand filter is measured. The coefficient of permeability k of the soil adjoining the filter is obtained from a graphical plot of the rate of flow against time (Wilkinson 1968; Fig. B.14).

The permeability is given by

$$k = \frac{Q_{(\infty)}\gamma_w}{Fa\Delta u}$$

where $Q_{(\infty)}$ is the volumetric flow rate extrapolated to infinite time, γ_w is the density of water, F is an intake factor depending on the ratio of the length to the diameter of the tip (for a sphere, $F = 4\pi$), a is the radius of the piezometer tip, and Δu is the increment in pore water pressure.

The coefficient of consolidation is given by

$$c = \left[\frac{Q_{(\infty)}a}{f\sqrt{\pi}}\right]^2$$

where f is the slope of the $Q(t)$ v. $\sqrt{t}$ line.

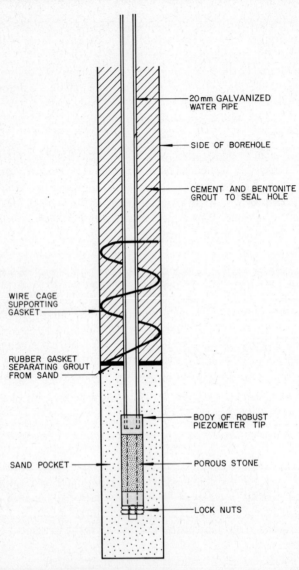

FIG. B.13. A piezometer for permeability measurements in a borehole and for monitoring fluctuations in the water table level: the tip is usually embedded in a sand pocket.

Open borehole tests most commonly measure permeability by introducing water into the hole at a rate that just maintains a constant water level. This rate measures lateral flow into the soil over the depth separating the water surface from the bottom of the hole. The coefficient of permeability is calculated in the same way as with a piezometer.

Well-pumping tests are often used to determine permeability in areas of shallow water table. At least five neighbouring boreholes must be sunk to intersect the horizon of interest. Water is pumped at a constant rate from the central borehole, while the water level is measured at regular intervals in the other holes. In highly permeable soils, water flows rapidly towards the central hole as its level is lowered, and this movement is reflected by a concomitant, though smaller, lowering of the level in nearby holes. The relationship between the water level in these holes and the pumping rate is used to calculate the average permeability of the intervening materials.

Ring-infiltrometer tests on the soil surface measure the vertical

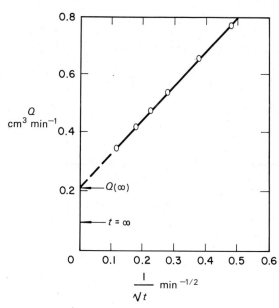

FIG. B.14. Typical results from a constant head permeability test.

permeability of the soil by introducing water into the soil through two concentric metal rings inserted a few centimetres into the uppermost soil horizon. The outer ring maintains the soil beneath it at the same moisture content as the soil beneath the inner ring, and so prevents water moving outwards from the column of soil beneath the inner ring. A more or less constant depth of water of about 50 mm is maintained in both rings by means of a float valve, which regulates inflow from a reservoir as the level drops. A cover protects the water surfaces from losses due to evaporation. The permeability of the soil is calculated from a graph of the rate of infiltration against time. Such a graph is useful in distinguishing the higher initial permeability of most dry soils from the lower sustained permeability of a soil with a higher moisture content.

The chief disadvantage of open borehole and well-pumping tests is the difficulty of relating their results to particular soil materials when there are several soil horizons between the water level and the bottom of the hole. This difficulty also affects ring infiltrometer tests, since the percolating water may pass through several different soil horizons. For this reason, piezometer tests are preferred when the permeabilities of several horizons must be distinguished. It should be noted that none of these field tests have reliable laboratory equivalents, owing to the small disturbances which inevitably occur during the recovery and trimming of an 'undisturbed' sample.

Modulus of elasticity, Poisson's ratio, modulus of subgrade reaction, and shear modulus

The modulus of elasticity, or Young's modulus, can be determined from the results of a *seismic survey* in terms of the following equation:

$$E = \gamma c_s^2 \frac{3 - 4(c_s/c_p)^2}{1 - (c_s/c_p)^2}$$

where E is the modulus of elasticity, γ is the bulk density of the soil, and c_s/c_p is the velocity ratio, or ratio of the velocity of S-waves to P-waves.

Poisson's ratio v is

$$v = \frac{0.5 - (c_s/c_p)^2}{1 - (c_s/c_p)^2}$$

(from Attewell and Farmer 1976).

Appendix D gives details of the procedures used in seismic surveys; note that to determine these moduli both the first (P-wave) and second (S-wave) returns must be detected, which is not usually possible over long distances. Reverse traverses should be used to check the results, particularly if inclined horizons are present.

Abbis (1979) has developed a method for determining the elastic modulus E in the field by measuring shear wave refraction during seismic surveys. The seismic velocity through a material is related to its dynamic stiffness, or shear modulus G, by the equation

$$G = \gamma c_s^2$$

where γ is the density of the soil and c_s is the shear wave velocity.

In practice shear waves are often obscured by compression waves, so that the measurement of the velocity is difficult. If, however, an enhancement seismograph is used and the direction of the seismic shock is reversed at source, and the reversal of the signal is also recorded, most of the interference will be cancelled out, while the recording of the required signal is enhanced.

To obtain the equivalent static shear modulus these values must be corrected with information on the damping properties of the ground. If Poisson's ratio v of the ground is known, Young's modulus may be calculated from $E = 2(1 + v)G$.

The modulus of elasticity, and the shear strength of a soil can be measured with a *pressuremeter* (Fig. B.15(a)), which measures the pressure required to expand a cavity in the soil. The most sophisticated of these devices is the self-boring pressuremeter, or Camkometer; this creates less disturbance than those types which must be driven into position and so displace the surrounding soil, or must be inserted into pre-drilled holes which cause release of pressures in the soil before the test. They can provide information to almost any depth, but their reliability is affected by irregularities such as thin soil layers or the presence of stones.

Although test procedures are fairly well established for saturated clays, the techniques of field installation and the interpretation of results are still being developed for sandy soils. The following very brief summary is based on the work of Wroth (Windle and Wroth 1977; Hughes, Wroth, and Windle 1977) who plots the results of an expansion test in a saturated clay as a graph of applied pressure ψ against radial strain ε (Fig. B.15(b)); a plot of shear stress can be derived from this by graphical analysis.

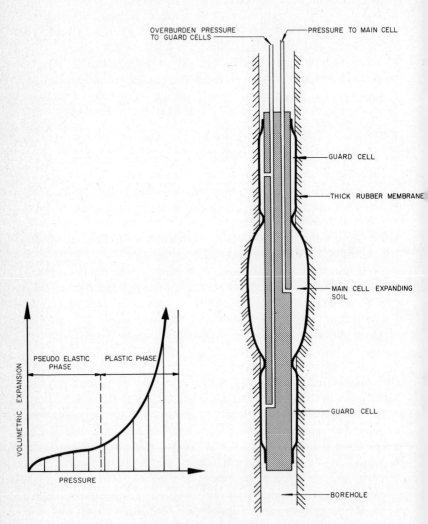

FIG. B.15(a). Principle of operation of the simple pressuremeter.

It is possible to obtain the following soil parameters from these results:

(i) The undrained shear strength $c_u = \tau_{max}$.

(ii) The shear modulus G from the slope of the line representing the reloading cycle:

$$\frac{\Delta\psi}{\Delta\varepsilon} = 2G.$$

From this a value of E_u may also be found:

$$E_u = 3G.$$

(iii) The *in situ* horizontal total stress σ_h is given by the value at which the radial expansion increases rapidly under pressure (point O in Fig. B.15(b)). Then

$$\sigma'_h = \sigma_h - u_0$$

where σ' is effective stress and u_0 is the initial pore pressure.

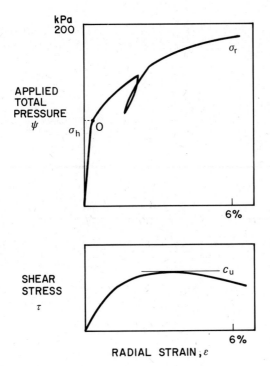

FIG. B.15(b). Results of an expansion test in clay using a Camkometer.

From this the value of the *in situ* earth pressure coefficient K_0 may be obtained:

$$K_0 = \frac{\sigma'_h}{\sigma'_v}$$

h refers to horizontal and v to vertical.

In sands the form of the pressuremeter curve is similar to that obtained from undrained tests in clay, except that a limiting pressure is not approached because of the volume changes occurring in the soil. A typical result is shown in Fig. B.15(c), in which

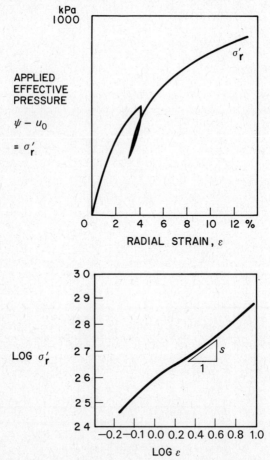

FIG. B.15(c). Results of an expansion test in sand using a Camkometer.

the effective radial pressure in the sand in contact with the membrane is plotted against the radial strain. When plotted on a double log scale the gradient s of the line thus obtained allows further interpretations of the results based on the theory of stress-dilatancy in sands (Rowe 1962, 1971).

The angle of shear resistance of the sand is given by

$$\sin \phi' = \frac{(K + 1)s}{(K - 1)s + 2}$$

where K is Rowe's stress dilatancy factor.

The angle of dilatation η may also be found from

$$\sin \eta = \frac{2Ks - (K - 1)}{K + 1}.$$

As is the case with clays, the shear modulus G can be obtained from the results of reloading cycles from

$$\frac{\Delta \psi}{\Delta \varepsilon} = 2G.$$

The lateral effective pressure may also be calculated, but its values are very sensitive to slight disturbance of the sand and should therefore be regarded as no better than approximate.

The shear modulus and the elastic modulus of the soil immediately below the ground surface or the bottom of an open excavation are best determined by a *plate-bearing* test. In this test a load is applied to a rigid plate in increments, at intervals of about 15 min or more, and the extent of the settlement under each increment is measured (Fig. B.16(a)). The plate must be properly bedded on the soil so that all parts of it are evenly in contact with the undisturbed soil surface.

In *clays* which are nearly saturated (i.e. in which no volume change will occur under quick loading) the shear modulus may be calculated from the slope of the curve relating stress to strain during a *reloading* cycle (Fig. B.16(b)) by the following equation:

$$G = \frac{q}{\rho} \frac{\pi D}{8} (1 - v)$$

where q is the average pressure on the plate, D is the diameter of the plate, and ρ is the settlement of the plate.

Young's modulus may be obtained from the shear modulus; in a saturated clay this is given by $E = 3G$. If the plate test is conducted at the bottom of a large diameter trial hole a depth factor must be applied to these results to allow for the stiffening effect of the overburden (at depths greater than $6D$, this depth factor is 0.85).

In addition, the ultimate strength of the soil can be determined by loading the plate until shear failure occurs beneath it.

The undrained shear strength of the clay c_u may be estimated from the ultimate bearing pressure on the plate using the relationship

$$c_u = \frac{q - \gamma Z}{N}$$

where γ is the bulk density of the soil, Z is the depth to the plate, and N is a bearing capacity factor which is often taken as 9.6.

The advantage of this test is that it provides a direct measure, under field conditions, of immediate settlement (i.e. the settlement which will occur when the structure is first loaded), as well as of shear strength. However, since the effects of the applied pressure are limited to depths of about 1.5 times the diameter of the plate, the

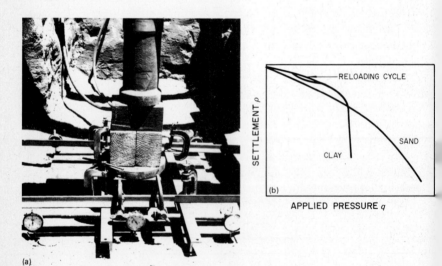

FIG. B.16(a). Load being applied during the execution of a plate-bearing test. (b). Typical results of a plate-bearing test.

test must be carried out from the bottom of an excavation or trial hole if the relevant properties of the *in situ* soil are to be established even to moderate depth.

In some soft or sandy soils it is possible to apply similar loads to *screw-plates* which are rotated into the soil to the required depth. Such a plate is akin to a portion of the flight of a spiral auger; the spiral has a flat pitch of about 20 to 25 per cent of the diameter (Fig. B.17). The results obtained from small plates may not be representative of those experienced when large areas are loaded, but are likely to be more representative than those obtained in the laboratory from small 'undisturbed' samples.

In *sands*, and soils in which drainage can occur during the loading test, the stress at which shear failure takes place is not clear from the stress–strain curve, and it requires engineering judgement to select the ultimate bearing capacity. In cases of doubt the ultimate capacity is often taken as that which causes a settlement of 15–20 per cent of the diameter of the plate. The shear strength of a sandy soil is not as easily estimated from its ultimate bearing capacity as the shear strength of clays, but if tests are carried out on various sizes of plate the results can be extrapolated to a larger foundation.

It is possible to estimate the likely settlement of a large foundation on a layered sandy profile from screw-plate tests carried out

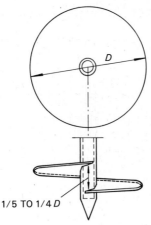

FIG. B.17. A screw-plate, similar to the flight of a spiral auger, may be used to carry out a plate-bearing test below the soil surface.

at various depths; the gradient of the stress–strain curve during a reloading cycle is measured at appropriate stress levels (Fig. B.16(b)). Then an equivalent modulus of compressibility E' is given by

$$E' = \frac{q}{\rho/D}.$$

If the increase in stress in each layer of the mass due to the foundation load is calculated from the theory of elasticity using a Boussinesq distribution of stress (Boussinesq 1885) then the appropriate modulus E' can be used for each level. The total predicted settlement is the sum of the separate estimates for each layer.

The plate-bearing test can also be interpreted to give the *modulus of subgrade reaction k* which is defined by the equation

$$k = \frac{q}{\rho}.$$

This modulus is useful in road and runway design which is based partly on theory and partly on empirical adjustments based on experience. Although the graph of settlement against depth is curved, a linear relationship is assumed and the slope is calculated for a specific point on the curve. In theory the modulus of subgrade reaction varies inversely with the diameter of the plate, so corrections are usually made to a diameter of 762 mm.

This modulus is also used when evaluating the effect of lateral loads on piles embedded in the soil; in this application the effect of plate size is again important. In both uses judgement is essential in selecting the practical range over which the results of the theoretical analysis may be applied.

The rate at which the structure is to be loaded, and the internal drainage of the soil, must be taken into account in assessing the relevance of results from plate-bearing tests: in impermeable soils the pore-water pressures will not dissipate sufficiently rapidly for the full settlement of the plate to take place.

When there is a compressible layer at shallow depth, piles are often used to transmit the load down to a deeper layer of soil or rock with a greater bearing capacity, or to utilize friction between the soil and the shaft of the pile. The predicted performance of a pile is often checked by a *pile-loading test*. The test load is applied by jacking the pile into the soil against a stable platform loaded with a kentledge of concrete blocks or iron weights to provide the necessary reaction; a mass of several hundred tonnes is sometimes required (Fig. B.18). Two types of test are commonly used. In the first, successive increments of load are maintained until the rate of settlement is very small; a lower limit of about 0.3 mm h^{-1} is often used (Fig. B.19(a)). This procedure is used to determine the ultimate bearing capacity of the material below the pile; if the actual point of failure is difficult to define this is often taken as that load which causes a settlement of 10 per cent of the pile diameter. The settlement under working conditions is assessed from the records of load, settlement, and time taken to reach equilibrium.

In the other method the pile is continuously loaded to achieve a constant rate of penetration of about 0.75 to 1.5 mm min^{-1} (Fig. B.19(b)). This procedure gives only ultimate bearing capacity, and

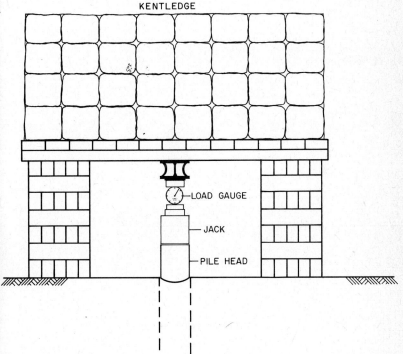

FIG. B.18. The ultimate bearing capacity of a soil at depth may be determined in the field by a pile-loading test.

no information is available on expected settlement. It is, however, a quicker test to perform. The shape of the load/penetration curve may reveal useful information on the mechanism through which sands resist the penetration of the pile; friction piles usually require less displacement to mobilize the full soil resistance than end-bearing piles.

In either test it is useful to unload and reload the pile to

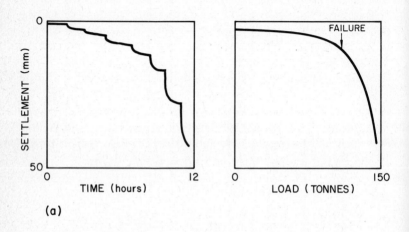

(a)

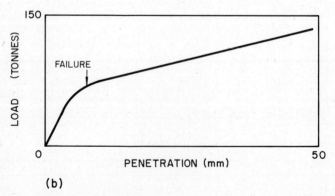

(b)

FIG. B.19. Typical results of pile-loading test: (a) with maintained load, (b) with constant rate of penetration.

ascertain the rebound characteristics of the soil. Since a full-scale pile test effectively duplicates most conditions that are likely to be experienced in practice, it can give valuable information on the performance of the final structure. However, it is not always possible to reproduce the actual rate of settlement, especially in clayey soils with high pore-water pressures; care should also be taken to ensure that conditions at the location of the test are representative of those occurring throughout the site.

Earth pressure

The pressure exerted by the soil on structures such as retaining walls can be measured with earth-pressure cells if these are carefully installed and properly calibrated for the soil type in which they are to be used. Any cell or gauge requires some deformation to activate the sensing element, and this can change the stress condition in the soil mass. The stiffness of the cell, its dimensions and temperature susceptibility are important factors in the design of the monitoring system. It is also important to know the behaviour of the ground-water during the period of measurement (Fig. B.20). This figure

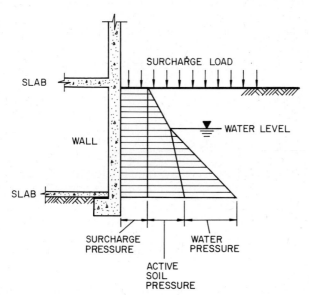

FIG. B.20. The distribution of lateral pressures behind a retaining wall.

also shows how the earth pressure can be doubled by the existence of a hydraulic head; drainage, either behind the wall or through weep-holes, helps to relieve such water pressures. In view of the difficulties encountered in obtaining reliable measurements the effort may only be warranted when information on the effects of a surcharge, or of compaction of the backfill, has major economic implications in a large project.

Shear strength

Some idea of the shear strength of both cohesive and non-cohesive soils can be obtained from the plate-bearing, screw-plate, and pile tests described above. The plate-bearing test only provides information to shallow depth; somewhat greater depths can be attained with screw-plates, while pile tests are effective to depths of 20 m or more, albeit at considerable cost.

The shear strength of most types of soil can also be determined by the pressuremeter tests described above (p. 301) in both shallow and deep investigations.

The strength of cohesive soils can be measured directly to considerable depth with a *Swedish vane*. The vane is housed in a protective 'torpedo' which is lowered into a borehole (Fig. B.21). The vane is then pushed out of the torpedo into the undisturbed soil below the bottom of the borehole, the measuring head is fixed to the outer casing of the vane and the casing is held stationary with a clamp. The torque required to rotate the vane at a uniform rate is recorded by the measuring head; the peak shear strength of the soil is calculated from the maximum value of the torque and the dimensions of the vane. The vane is usually rotated by hand through a gear-wheel at a rate of about 0.1 degrees s^{-1}, which is slow enough not to generate any viscous resistance, but is still fast enough not to allow dissipation of pore-water pressures, i.e. the clay is tested under undrained conditions. Tests can be conducted under drained conditions if the vane is rotated at a slower rate of 0.1 degrees min^{-1}, but this procedure is time-consuming and relies on a motorized version of the apparatus. If the ratio of the height of the vanes to their width is varied during successive strength determinations in the same soil an estimate of the ratio of the shear strengths in the horizontal and vertical planes can be obtained. This can, in turn, be interpreted to provide an indication of the *in situ* earth pressure coefficient (Blight 1970).

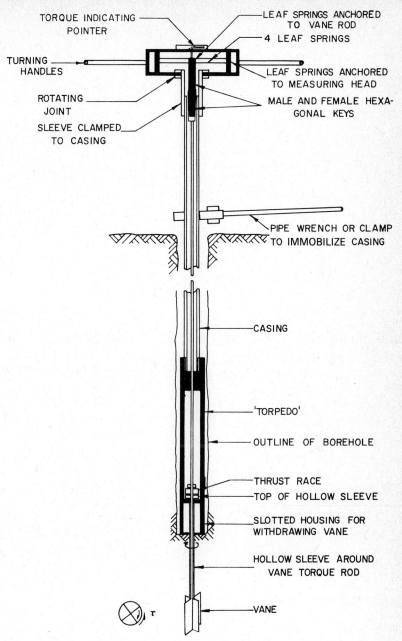

FIG. B.21. Swedish vane apparatus for the *in situ* measurement of the shear strength of clayey soils.

The equation relating torque to shear strength is

$$T = \frac{\pi D^2}{2} \tau \left(H + \frac{D}{3} \right)$$

where T is the torque, D the diameter of the blades, H the height of blades, and τ the shear strength of the soil. If $H = 2D$ as is usual, then

$$\tau = \frac{6T}{7\pi D^3}.$$

After shear, the vane is rotated several times to remould the soil thoroughly, and after a short delay the resistance of the remoulded material is measured in the same way to give the residual shear strength of the soil.

This test is restricted to cohesive soils; its reliability is considerably reduced by irregularities such as stones and laminations in the soil.

The Dutch probe can be used to give a guide to the *in situ* shear strength of both *cohesive* and *non-cohesive* soils in deep investigations. This test was devised in Holland to facilitate the design of piles. As is the case with other empirical tests, such as the standard penetration test described below, the results can be correlated in a general way with various soil properties, provided that the test is not used out of the context for which it was devised. The equipment comprises a 60° cone with a cross-sectional area of 100 mm² which is pushed into the soil below the bottom of a borehole with a hydraulic ram. Several types of cone are available, but the friction-sleeve cone shown in Fig. B.22 is commonly used. This sleeve provides a direct measurement of the friction which can be generated when a pile is driven into the soil. The pressure required to advance the cone and the friction between the sleeve and the soil are measured every 200 to 250 mm over the depth of the sounding, which is normally taken to some more resistant horizon or to bedrock.†

---

† In a recent development, a continuous record of pore-water pressure can also be obtained as the soil is penetrated; if penetration is halted and the excess pore pressure is allowed to dissipate, the level of the water table can be determined.

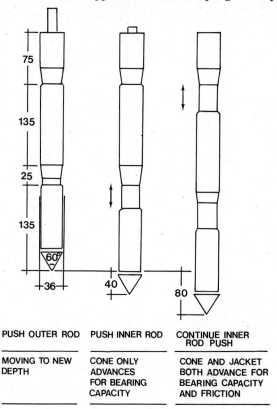

PUSH OUTER ROD	PUSH INNER ROD	CONTINUE INNER ROD PUSH
MOVING TO NEW DEPTH	CONE ONLY ADVANCES FOR BEARING CAPACITY	CONE AND JACKET BOTH ADVANCE FOR BEARING CAPACITY AND FRICTION

FIG. B.22. Mechanical friction-sleeve cone for Dutch probe test (dimensions in millimetres).

The relationship between the cone resistance q_c and the side friction f_s is expressed as the friction ratio R_f, i.e.

$$R_f = f_s/q_c \times 100 \text{ per cent.}$$

This ratio gives an indication of the type of material being penetrated, low values of about 3 per cent indicating sandy soils and higher values of between 4 and 8 per cent indicating clayey soils.

Fig. B.23 shows typical graphs of the change with depth of resistance and friction as measured with a Dutch probe. These

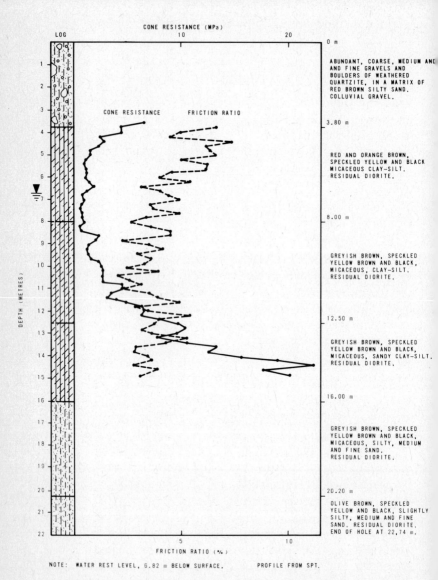

FIG. B.23. Results of Dutch probe test. The soil profile was recorded from disturbed soil samples recovered with a Raymond spoon in a nearby borehole.

measurements can be used to calculate the allowable bearing pressure on the soil beneath piles.

Fugro's (1974) formula has been widely used in sands:

$$q_{\text{allow}} F = 0.25 \, q_{c0} + 0.25 \, q_{c1} + 0.5 \, q_{c2}$$

where q_{allow} is the allowable pile-bearing pressure, F is a factor of safety (usually taken as 2 for cylindrical piles), q_{c0} is the average cone resistance over 2 pile diameters below the base of the pile, q_{c1} is the minimum cone resistance over 2 pile diameters below the base of the pile, and q_{c2} is the average of the minimum cone resistance over 8 pile diameters above the base of the pile.

In cohesive soils, the cohesion c_u of the material is given by the relationship

$$c_u = \frac{q_c}{n}$$

where n is a factor which varies from 9 to more than 25, depending on the stiffness of the soil. The adhesion of the soil to the pile shaft is calculated by multiplying the cohesion by a factor varying from 0.3 to 1.0, again depending on the stiffness of the soil and the method of installation of the pile. The allowable bearing pressure q_{allow} is obtained from

$$q_{\text{allow}} = \frac{q_c}{3}.$$

The computation of pile or footing settlements from cone resistance and friction ratio is more complex; for a detailed treatment the reader is referred to Schmertmann (1970) and Jones (1978).

Like the standard penetration test, the Dutch probe test can be carried out rapidly and relatively inexpensively in difficult terrain (even in marshy estuaries) using relatively light equipment. Its results are generally considered more reliable than those from the standard penetration test, but soil samples are not usually recovered during a Dutch probe test, and so the calibration of the results against soil conditions requires stratigraphic information from other sources.

The resistance of *non-cohesive* soils to penetration is determined either with a Dutch probe or by the *standard penetration test* (SPT). Both can be carried out to considerable depth in boreholes (up to 50 m or more if required), but in saturated sands the sides of the hole must always be supported by casing. The SPT is the most widely used of all *in situ* tests. A standard split-spoon sampler, or Raymond spoon (Fig. B.10), is driven into the undisturbed soil below the bottom of the hole by repeated blows of a 63.5 kg mass falling through 762 mm. The split-spoon sampler is driven through a distance of 457 mm and the number of blows needed to penetrate the last 305 mm is known as the standard penetration number or *N*-value. After the *N*-value has been established, the split-spoon is withdrawn from the borehole and opened to extract a disturbed sample of the soil which was penetrated. This procedure is repeated at selected intervals, between which the hole is deepened by boring. The soil profile can be roughly reconstructed from these disturbed samples. Fig. B.24 shows a typical soil profile and a graph of the *N*-values from the results of a standard penetration test.

The *N*-values of a saturated soil can be related in a general way to its cohesion (in a clayey soil) or its relative density† (in a sandy soil); these, in turn, govern its consistency. Table B.3 presents relationships between the *N*-value and these parameters. It should be noted that the correlations are less reliable for cohesive soils than for sands. Several correlations between *N*-values and the allowable bearing pressures on foundations have been proposed; Fig. B.25 illustrates such a correlation for footings of different width in sand, while Fig. B.26 relates allowable bearing pressure to *N*-values for a number of soils, including some which show cohesion. Both correlations assume that a total settlement of about 25 mm will be acceptable. The *N*-value can also be used to estimate the likely settlement of foundations at different depths under various loads (Sanglerat 1972). However, all such estimates of allowable pressure or of settlement from *N*-values should be viewed with caution and, if critical, should be checked by other methods.

Like the Dutch probe, the SPT can be carried out quickly and at relatively small cost in difficult terrain because the equipment is

---

† The *relative density* (RD) of a sand is the ratio (expressed as a percentage) of the existing state of compaction to the loosest and densest states it can achieve:

$$RD = \frac{e_{max} - e}{e_{max} - e_{min}} \times 100$$

where e is the present void ratio, e_{max} is the void ratio in the loosest state, and e_{min} is the void ratio in the densest state. The maximum and minimum void ratios can be determined by laboratory procedures.

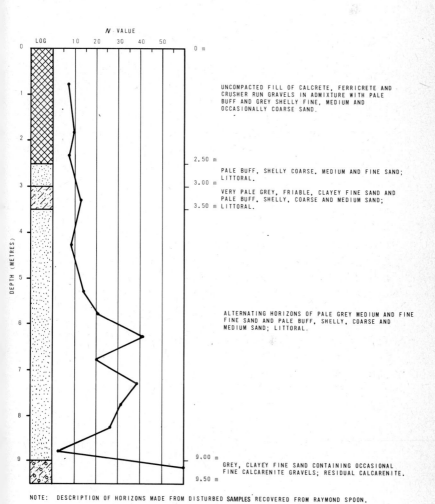

FIG. B.24. Results of a standard pentration test (SPT).

easily portable. It must be clearly understood that the interpretation of soil parameters from the results of either of these tests is based on empirical relationships derived from experience. It is essential that the established or standard procedure of testing is strictly adhered to, or the results may be meaningless. In all cases the standard to be used should be discussed with the engineer directing the work before testing is undertaken.

Also useful in a variety of soils is the *dropweight cone penetrometer* (DCP), in which a cone with a diameter of 20 mm is driven into the soil by a 7.815 kg weight dropped through 575 mm (Fig. B.27). The results are expressed as millimetres penetrated per blow. A modification provides a slot in the rods to permit the

TABLE B.3

Results of simple probing tests can be related to soil consistency, the relative density of sands and the cohesion of clays. (Partly after Terzaghi and Peck 1967)

	Sandy materials		
Description	Burmister's relative density (%)	Saturated standard penetration test N (blows per 300 mm)	Dropweight cone penetrometer (mm per blow)
Very loose	—	<4	>75
Loose	0–40	4–10	30–75
Medium dense	40–70	10–30	12.5–30
Dense	70–90	30–50	5–12.5
Very dense	90–100	>50	2–5

	Clayey materials		
Description	Cohesion (kPa)	Saturated standard penetration test N (blows per 300 mm)	Dropweight cone penetrometer (mm per blow)
Very soft	<18	<2	>110
Soft	18–36	2–4	55–110
Firm	36–72	4–8	30–55
Stiff	72–144	8–15	15–30
Very stiff	>144	15–30	7–15

recovery of a small, disturbed sample of the soil. The light, portable equipment used in this test limits its range to a depth of 1 to 3 m below surface. It is, therefore, most useful for estimating soil conditions during the design of shallow footings or for assessing subgrade soils for road design. However, the ease and low cost with which results can be obtained is somewhat offset by the indirect approximation to soil conditions that it provides. Table B.3 gives some relationships between penetration and soil consistency, the relative density of sands, and the cohesion of clayey soils.

A crude approximation to the consistency and strength of *in situ* soils may also be achieved by calibrating the *rate of penetration* of rotary percussion or core drills with the properties of undisturbed samples from the same materials. This can provide useful information during deep investigations in a wide variety of soils when no better means of estimating the soil properties are available.

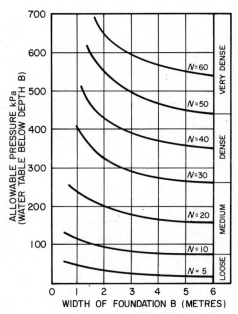

FIG. B.25. Chart for estimating the allowable bearing pressure for foundations in sand from the results of a standard penetration test. *N*-values are the number of blows required for 300 mm penetration. It is assumed that 25 mm of settlement can be tolerated. (After Terzhagi and Peck 1967.)

The relative merits of the various laboratory and *in situ* tests depend very much on circumstances. However, it is widely accepted that, while boundary and stress conditions can be accurately controlled in the laboratory, the soil samples recovered for laboratory analysis are often somewhat disturbed; in fact, it is almost impossible to obtain undisturbed samples from sandy and silty soils with little cohesion. Furthermore, small laboratory samples are not always representative of that part of the soil mass with which the engineer is concerned.

The properties of sands are best revealed by *in situ* tests, of which the most reliable is the plate-bearing test, followed by the Dutch probe and, finally, by the standard penetration test. In cohesive soils the SPT is of dubious value, but the Dutch probe often gives useful results if the cone resistance is carefully correlated with rate of penetration and shear strength in the local soils involved. The Swedish vane offers the most accurate results in these clayey soils, but pressuremeters may prove useful if difficulties over the interpretation of its results can be overcome.

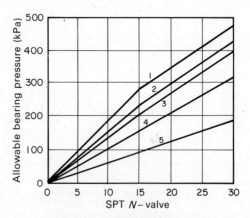

1. PARTLY SATURATED CLAY, NEVER INUNDATED
2. SATURATED CLAY, SQUARE FOUNDATIONS.
3. DRY SANDS, FINE GRAVEL
4. SATURATED CLAY, LONG FOUNDATIONS;
   FLOODED SANDS AND FINE GRAVEL.
5. ORGANIC SILTS AND ORGANIC CLAYS
   WITHOUT PEAT.

Fig. B.26. Relation of allowable bearing pressure to N-value for various soils under column loads ranging from 100 to 450 kPa. (After Leonards 1962.)

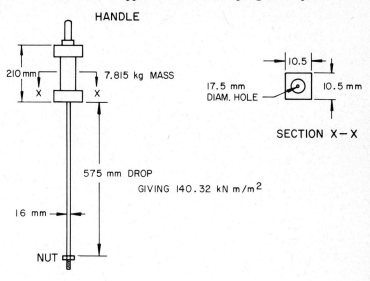

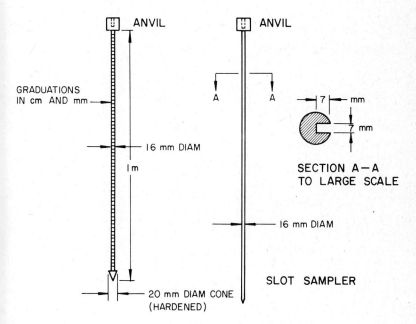

FIG. B.27. Dropweight cone penetrometer (DCP) for estimating soil consistency to shallow depth.

Appendix C. Remote sensing techniques

Remote sensing imagery is reproduced from data gathered by sensors (e.g. cameras, sensitive cells, or radio receivers) on remote platforms such as aircraft and satellites, which record radiation that has been emitted or reflected from the earth's surface in various electromagnetic wavebands (White 1977).

The chief constraints on the choice of platform are height and speed of transit. The former governs resolution, or the size of the smallest object which can be distinguished, while the latter determines the minimum height above the earth's surface from which a good image can be obtained with optical systems. With modern aircraft and instrument mounting systems, vibration seldom imposes serious constraints. In practice, imagery can now be obtained from a wide range of altitudes; satellite platforms provide wide coverage at small scales and low levels of resolution, intermediate scales and resolution can be obtained with aircraft, while large scales and high resolution are possible using helicopters and balloons.

Different types of imagery are distinguished according to the waveband within which they record electromagnetic radiation, on whether it was reflected (e.g. solar) or auto-emitted (earth-generated) radiation, and on whether the sensor system was passive or active. Passive systems, such as cameras and scanners, merely intercept and record radiation escaping from the earth, while active systems (e.g. radar) generate a signal which is reflected from the earth's surface.

Only microwave radiation (wavelength 1 mm to 3 m) suffers so little attenuation through the earth's atmosphere that it may be used in the active mode. Sidelooking radar (SLR) operates within the 7 mm to 30 cm wavebands and the intensity of the return beam is determined chiefly by surface micro-relief and variations in the moisture content of soil and vegetation. One traverse covers a path up to 40 km in width, but the resolution is rather poor (> 10 m) at the useful operating scales between 1:50 000 and 1:500 000. The chief application of SLR is in penetrating cloud cover to produce a photograph-like image of the land surface (White 1977). This facility and the small-scale imagery and low resolution make SLR most suitable for regional surveys in the cloud-covered tropics.

Although commercial systems have not yet been developed, lasers may offer an active mode in other wavelengths.

Some radiation, particularly in the infrared and microwave bands (>780 nm and >1 mm wavelengths, respectively), is emitted directly by the earth's surface and received passively.

Thermal infrared radiation in the wavebands 3000–4000 nm must have originated at the earth's surface and at shallow depths below it (less than a few centimetres), since radiation from greater depths is severely attenuated. The use of thermal infrared imagery depends on its capacity for distinguishing materials with different thermal properties, which are related to soil texture and mineralogy and the ability of the material to retain moisture, and so the most useful images are obtained from pre-dawn missions when reflected radiation is absent, and sufficiently long after heating has ceased to show contrasts in thermal inertia. Since this waveband is beyond the limit of the visible spectrum (780 nm) and cannot all be focused simultaneously onto a recording medium such as film, the signals are collected in the 'object plane' by means of a scanning detector which builds up the image on magnetic tape in a series of 'slices'. The results are converted into analogous light signals for recording on photographic film, on which by common convention hot areas are represented by light tones and cold areas by dark tones. Attenuation of the signal between the radiating surface and the sensor is exacerbated by microclimatic effects such as cold and warm air currents, by high soil moisture content and by dense vegetation.

Radiation emitted by the earth in the microwave band is usually recorded by scanners operating at wavelengths of 1.0 to 30 mm. These systems receive radiation from considerably greater depths (many metres below the surface) than infra-red linescanners, but the signals are severely attenuated by soil moisture and provide poor resolution. Their chief application is in arid areas.

The remaining procedures record reflected solar radiation. Solar radiation and terrestrial reflectance are highest in the visible wavebands (380–780 nm) but certain wavelengths are filtered out by the atmosphere. Notable atmospheric 'windows' occur in the near to middle infrared wavebands (3000–14 000 nm), and in the microwave band and beyond (> 1 mm). The image intensity varies considerably with the reflectance of the objects illuminated. The proportions of the different wavelengths that are reflected by an object are its 'spectral signature', which depends not only on its colour but also on its surface roughness and moisture content which cause shifts to longer wavelengths. The reflected radiation may be recorded in terms of the three primary colours (as in colour photography) or in finer subdivisions (multispectral scanning) or in tones of grey (as in panchromatic photography where the nature of the particular film and filter determines the weighting given to different wavelengths). The degree to which the reflected radiation received by the sensor differs from that recorded by a spectro-radiometer near the reflecting object measures the attenuation between object and sensor. Radiation in the visible wavebands is severely attenuated by water vapour and dust in the atmosphere but gives the best resolution under good conditions. Since reflected radiation from all of the visible waveband can be focused simultaneously onto the recording medium the sensor system is said to record on the 'image plane'.

Such reflected radiation is usually recorded on photographic film. Panchromatic and colour films are sensitive to the full visible waveband (380–780 nm), while infrared films (both panchromatic and colour) are sensitive also to the invisible near infrared spectrum (780–900 nm), which increases their discrimination of roughness and moisture contrasts. Contrasts are sometimes accentuated by weighting specific wavebands, through special films or by filters. In multispectral scanning the reflected image is divided into several waveband components by filters or by using a battery of recording sensors, each recording a different waveband in the object plane. Multispectral images are most useful when the spectral signatures of features of particular interest are already known; for example, target analysis on the ground with a spectro-radiometer may indicate which wavebands or filters are most discriminating.

Photographic images may be converted to digital form by scanning, and the spectral data for every small area of the image (pixel) are then stored (e.g. on magnetic tape) for transmission or manipulation. Where scanners are used to record spectral data in the object plane, direct storage on magnetic tape is normally provided (e.g. in most contemporary satellite systems).

MAPPING FROM REMOTE SENSING IMAGERY

Imagery obtained from remote platforms has two main roles in soil mapping; to *match* unknown to known sites by comparing their spectral signatures, or to *recognize* the nature of the emitting or reflecting surface directly from its spectral data.

In either case the image is scanned to produce digital data. Spectral signatures can then be classified and matched by computer. Automatic signature recognition involves the numerical identification and classification of spectral signatures according to their density ranges in one or several wavebands. Ranges of signatures may be determined for the corresponding ground features using a spectro-radiometer in the field. Image and ground data may then be matched by techniques such as clustering algorithms, likelihood ratio algorithms or table comparisons (Van Vleck, Sinclair, Pitts, and Slye 1973).

Computer mapping from spectral imagery by means of signature matching has shown some promise in mapping land use and vegetation in local areas, but has not been very successful in mapping soil types, for which there is, as yet, no substitute for human interpretation. This is also essential to the second role of remote sensing imagery, which uses photographic images, either of data in the visible spectrum which was recorded directly on the image plane, or of scanner information reproduced on photographs. Conventional aerial photographs record data in the image plane and have the advantage that, when successive images are recorded along the flight path of the aircraft so as to provide a substantial measure of overlap (usually not less than 60 per cent), they create a three-dimensional model when viewed through a stereoscope. Table C.1 lists some relationships between camera, flying altitude, scale, and coverage of aerial photographs. As is the case with automatic matching, the interpreter groups sites on their spectral signatures (e.g. tone or colour value), but he also relies on his own ability to *recognize* the objects which he observes. This is aided by his experience in interpretation, especially in discerning subtle spatial changes, in evaluating likely correlations between different classes of natural features, and in eliminating the superimposed manifestations of man's activities.

Multi-spectral images, or colour or false colour photographs produced by dividing the spectrum reaching the film into different

TABLE C.1

Some data on flying height and airphoto coverage for selected photo scales

Photo scale (230 × 230 mm format)	Flying height (m)	Coverage of photo (m)	Flight line separation (m)		Forward gain with 60% overlap (m)	Number of prints per line kilometre	Number of strips per kilometre width		
			30% sidelap	25% sidelap			30% sidelap	25% sidelap	
1:1000	152	230	160	170	91	10.938	6.246	5.836	152.4 mm focal length camera
1:10 000	1524	2300	1600	1710	914	1.044	0.628	0.584	
1:20 000	3048	4570	3200	3430	1830	0.547	0.311	0.292	
1:30 000	4572	6860	4800	5140	2740	0.367	0.208	0.195	
1:40 000	6096	9140	6400	6860	3660	0.273	0.156	0.146	
1:50 000	7620	11 430	8000	8570	4570	0.219	0.125	0.117	
1:50 000	4418	11 430	8000	8570	4570	0.219	0.125	0.117	RC9 super wide-angle camera (focal length 88.36 mm)
1:60 000	5302	13 700	9600	10 290	5485	0.182	0.104	0.097	
1:70 000	6185	16 000	11 200	12 000	6400	0.156	0.089	0.083	
1:80 000	7069	18 300	12 800	13 715	7315	0.137	0.078	0.073	
1:90 000	7952	20 575	14 400	15 430	8230	0.122	0.070	0.065	
1:100 000	8836	22 860	16 000	17 145	9145	0.109	0.063	0.058	
1:150 000	13 254	34 290	24 000	25 720	13 715	0.073	0.042	0.039	

components by means of colour filters, may provide additional aids to such recognition. In addition the image may be 'enhanced' to give prominence to particular features at the expense of other data. Thus particular contrasts in photographic images may be enhanced by the preferential reproduction of particular levels of film density (density slicing). This is particularly useful to accentuate the boundaries of tonal units in the photographs (edge enhancement), or it may be used to highlight transitional zones that represent particular changes in soil properties. Density slicing of band 7 of Landsat imagery has been shown to enhance the appearance of gravel occurrences, pedocrete crusts and rock outcrops (Beaumont 1977), while other slices may accentuate differences in moisture and vegetation (Anuta, Kristof, Levandowski, Phillips, and MacDonald 1971). The same process of selecting signal levels corresponding to specific ranges of response in the image from the photograph, or from data recorded directly on magnetic tape, may be applied to video images, though its usefulness is somewhat reduced by the poor resolution inherent in the low line densities available in most contemporary television systems.

Contrast stretching, achieved by the use of high-gamma film or logetronic printing may be used to enhance low contrasts in the image that correspond to differences in soil types. Special super-imposition techniques can produce similar results. For example, image masking by superimposing images from different wavebands, often with addition of appropriate colour elements by means of filtering, provides colour composite images that highlight specific features. Image combination can also be achieved with colour additive viewers, which superimpose images in different wavebands by projecting them simultaneously onto a viewing screen: each image is projected in a different colour and the brightness of the different images is adjusted to give special prominence to a desired feature. Similar results can be produced by the selective printing of filtered images from separate wavebands. Such manipulations are useful for distinguishing textural classes in soils and for giving better definition to areas of slope instability.

Similar enhancement can be carried out by digital computer techniques if the spectral information is available in the form of single, spatially defined elements (pixels).

INTERPRETATION OF REMOTE SENSING IMAGERY

The interpretation of remote sensing imagery for soil survey aims to relate particular patterns on the imagery to particular soil profiles—i.e. the form (or image) must be related to the substance (or profile). Once a pattern (e.g. displayed by natural vegetation) has been identified on one part of an image, an intelligent observer may recognize the same pattern elsewhere on the image; further study, aided by stereoscopic viewing to show relief, often reveals that both patterns are associated with the same kind of landform (e.g. remnants of a river terrace). If a similar soil profile is present beneath the surface of all remnants of the terrace within a given area, then areas on the image with the same pattern and the same landform may be delineated to give a tentative map unit. This sequence of recognition and correlation is fundamental to the effective use of remote sensing imagery in the preparation of soil maps.

Successful interpretation depends largely on the experience of the interpreter; his expertise is based on consciously or subconsciously accumulated criteria for correlating image signatures with field conditions. It is easier to acquire this talent by example than from reading. It must be clearly understood that soil boundaries annotated on remote sensing imagery cannot offer more than approximations to field conditions, and they cannot replace fieldwork. However, there is never enough time or money to achieve a sufficient density of field observations, and interpretations from imagery are often very helpful in approximating the position of the soil boundary already known to occur between field observation points. Annotations made in the early phases of a project will usually be cruder than those which become possible after lengthy familiarization with the area, but will none the less serve as a useful framework within which to plan the field programme.

To interpret the various types of aerial photography it is essential to obtain a stereoscopic image which shows the land surface in relief, usually with some degree of vertical exaggeration due to the optical characteristics of the stereoscope through which viewing is carried out. A small, folding, pocket stereoscope may be used to examine standard aerial photographs in the field. A larger mirror instrument, with which photographs can be fixed onto a movable base plate under parallel guidance system and with a built-in cold tube light source, is preferable for the laboratory. A binocular attachment of high magnification for detailed scanning is useful in large-scale mapping. The photographs may be annotated on glass-clear overlays or directly with coloured, erasable wax pencils.

It must be stressed that soil types cannot be *identified* directly from remote sensing imagery. The tone of the image results from variations in surface reflectance or emission. In arid and semi-arid environments a large proportion of the image may represent the exposed surface of the soil or rock. In general, smooth surfaces and fine soil textures (e.g. uncultivated or clayey soils) tend to produce paler photographic tones than rough surfaces or coarse superficial materials (e.g. cultivated areas or gravels). Moist soil materials also tend to have darker tones and, in any particular area, may indicate either a high water table or poor internal drainage due to a relatively impermeable subsoil horizon. These contrasts are often accentuated in infrared photography. Vegetation tones are mostly lighter in panchromatic infrared prints. In less arid and more heavily vegetated areas the ground surface itself usually occupies only a small part of the total image, and its main variations are due to vegetation. Changes due to human activities (e.g. agricultural features, settlements, dams, and communication lines) are super-imposed upon these features. Such cultural elements are often sufficiently large and regular to permit rapid identification, and the more important elements of vegetation and hydrology can frequently be recognized by analogy to known sites.

However, what the engineer usually wants to know about soils is the sequence of subsurface horizons and their chemical or physical properties, and different types of soil profile can very seldom be recognized directly from remote sensing imagery. It is seldom possible to attribute variations in surface reflectance or in tone or form to subsurface conditions without careful field checking. More commonly, soil changes are indirectly inferred from associated changes in natural vegetation, often in combination with correlated variations in micro-relief and landform. Changes in land use sometimes correspond to different soil properties. It is clear, therefore, that the delineation of soil units from remote sensing imagery is usually based on indirect evidence. In this lies a danger: the interpreter must not exploit correlations deterministically, without allowing for the independent variation of individual elements within the landscape, and must bear this risk in mind during field checking. It must also be remembered that the properties and distribution of soils can only be fully comprehended in the context of the landscape of which they form an integral part. Thus in a given area there is usually a characteristic vertical and lateral sequence of soil types, usually (but not invariably) associated with specific landforms and plant associations; recognition of the local catena is a most useful aid in soil mapping from remote sensing data.

The interpretation of remote sensing imagery is best undertaken in a series of successive steps, in each of which various elements of the image are considered separately. It may be useful to tabulate the appropriate items for each mapping unit in the manner suggested in the table accompanying Plate 11, where standard panchromatic aerial photography has been interpreted in terms of soil units defined on their origins and textures. The chief elements of each image are:

Tone, or the light reflectance characteristic of the surface.
Texture, or the pattern of variation of tone.
Shape and internal pattern, which can be divided into
 (i) *shape and structural pattern*, which define the outline form of the unit and the pattern of internal contrasts that relate to its geological structure;
 (ii) *drainage form*, which defines the configuration of hydrological features including their pattern, density and degree of integration.
Stereoscopic appearance, or the relief, slope and three-dimensional form of the unit.
Associative characteristics such as special features of land-use, soil erosion, vegetation type, etc.

These elements may be further analysed and interpreted to *recognize* soil type. For example, variations of tone and texture can often be resolved into elements produced by:

 (i) natural vegetation—for example the lines of thorn trees which may be present along dykes (d in Plate 6 or RD in Plate 11) or the corridor of high bush which may occur along an alluvial floodplain (A_E in Plate 5);
 (ii) surface soils or rock outcrops—for example the slightly raised greyish outcrop included in unit RO in Plate 8 or the bouldery ridge of unit 2 in Plate 4; except in arid and freshly ploughed areas, surface soils are usually covered by vegetation;
 (iii) land use—this usually obscures soil changes by superimposing cultural features such as fields, hedgerows, and roads upon the natural patterns; only where a close correlation exists between soils and land

use, as is the case in certain areas of Holland where crops are selected strictly according to soils, can such cultural features serve as more than a rough guide to soil changes.

Shape and structural pattern can be resolved into:

(i) plan form—for example the pear-shaped or hemi-lemniscate form of the gully heads (unit GWg/F/WG) in Plate 11;

(ii) structural pattern—for example the alternating bands of paler and darker tone which indicate different strata in dissected areas underlain by well-bedded sedimentary rocks.

Drainage form can be subdivided into:

(i) pattern—for example the trellis-like configuration imposed upon streams by the occurrence of a dipping sequence of resistant and soft horizons in Plate 9; this element is most useful for delineating large units at small scales (e.g. on satellite images);

(ii) density—for example the dense network of rills which is often associated with impermeable soil, in contrast to the less frequent stream channels which are evident in an adjoining area of permeable, sandy soil;

(iii) integration—for example the well-integrated system of continuous stream channels evident in Plate 6, where insoluble rocks and relatively impermeable soils favour surface runoff; poorly integrated systems occur often in karst and pseudokarst areas or where the drainage pattern has been disturbed by glaciation.

Stereoscopic appearance can be resolved into:

(i) landform, which is often the most important element on which different soils and rocks are distinguished, for example the shallow gullies containing active clayey soils in unit exp C of Plate 6, or dunes composed of windblown sand; the boundaries shown in Plate 7(a) are based almost entirely on the recognition of the different landforms produced by colluvial and alluvial processes operating in a sub-humid area underlain by a dipping succession of hard and soft rocks;

(ii) erosion form, or the cross-sectional profile of gullies eroded into the soil along the headwaters of streams; while seldom visible at scales smaller than 1:10 000, these cross-sections often permit broad generalizations about the cohesion and angle of shear resistance of the soil: a V-shaped cross-section suggests the presence of a cohesionless sand, while a rounded concave section is common in clays with a high cohesion (c) and low angle of shear resistance (ϕ), and a rectangular section with vertical sides usually indicates the presence both of cohesion and a moderate angle of internal friction, as is the case with most sandy clays and silts; in Plate 6 the narrow, steep-sided gully is cut into clayey transported soil.

Associative characteristics include such features as quarries for road gravel, the development of extensive badlands due to the absence of suitable conservation measures after ploughing dispersive clayey soils, and tall growth of reeds in swampy areas.

In interpreting colour imagery, *spectral value* should be substituted for tone, and is described in terms of combinations of the three primary colours. Some elements, particularly landform and erosion form, cannot always be identified if the imagery is not susceptible to stereoscopic viewing.

While analysing the elements of the image, the interpreter should not lose sight of the following controlling factors:

  (i) type of parent material: this will often control the landform and drainage pattern (e.g. rolling landscape with dendritic drainage pattern on granite–gneiss; hummocky landscape with deranged drainage pattern on recent till);

 (ii) geomorphological processes at work (e.g. the development of unstable slopes on deep residual soils in areas of high relief, or of sinkholes and dolines where karst solution is operative);

(iii) climate and physical environment (see Table 3.5, p. 98);

(iv) biotic environment: e.g. patterns produced by termite mounds are confined to specific soils;

 (v) human activities: landforms and drainage patterns modified mainly as a result of agricultural or engineering works, both civil and mining (e.g. the use of tiled drainage in cultivating waterlogged areas, and disturbances produced by strip mining).

Special procedures and precautions are necessary in the interpretation of multi-spectral images and infrared linescan imagery. In the former, it is usually necessary to compare several different waveband and filter combinations before the best contrasts are obtained. The combination should be recorded when tabulating the elements of the image of each mapping unit. Infrared linescan imagery cannot be reproduced in a form suitable for stereoscopic viewing, owing to the continuous scanning procedure and the large distortions present along the edges of the scan paths, so this type of imagery is best used in conjunction with conventional aerial photography. Even so special care should be taken to identify any extraneous effects produced by local atmospheric contrasts such as cold air pockets and warm winds, or by surface moisture after rain. Under good conditions, fine-grained relatively impermeable soil material with relatively low emissivities tends to present darker tones on infrared presented in the conventional photographic format than coarser, more permeable soils. Likewise, basic rocks and their derivatives tend to show darker tones than their more acid counterparts. Areas of surface moisture stress, resulting from good subsurface drainage, generally tend to display lighter tones.

COMPARISON OF DIFFERENT TYPES OF REMOTE SENSING IMAGERY

The relative advantages of different types of remote sensing imagery in soil mapping depend on the nature of the terrain, the scale of the imagery and the extent to which local effects have attenuated or confused the signal. In general, satellite imagery (e.g. that available from the current Landsat programmes initiated by the United States) and microwave imagery (e.g. side-looking radar) are suitable only for large regional surveys where the detail of soil information required is relatively low, and which exploit the relationships between soils, vegetation and landforms (e.g. in defining Land Systems, pp. 214–20). Landsat imagery is available from a multi-spectral scanner, operating in four channels covering the waveband 500–1100 nm, and from a vidicon camera operating in the waveband 440–840 nm (White 1977). The repetitive coverage permits the choice of whatever imagery displays the maximum seasonal contrasts related to soil moisture, but acceptable resolution cannot be achieved at scales much in excess of 1:250 000. By careful manipulation, stereoscopic or pseudo-stereoscopic parallax can be obtained from Landsat imagery, by using the 10 per cent overlap between scan paths, by simultaneously observing images from different channels, or by exploiting the shadow parallax present in coverage of the same area obtained during different seasons. A great advantage of Landsat imagery is its synoptic view, which facilitates the identification of relationships which are not apparent in imagery giving restricted coverage at larger scales.

The utility of infrared linescan imagery may sometimes be impaired by its inability to produce a stereoscopic image, by its inherent distortion along the edge of scanning strips, by the masking effects of cold air pockets, warm winds and vegetation, by the blurring of contrasts by high soil moisture contents, and by the overlapping of the thermal properties of superimposed soil materials. These difficulties are not as formidable as they appear, and good images can be obtained under favourable weather conditions provided that the mission is not undertaken too soon after rain. It is likely to give the best results in sparsely vegetated arid to sub-humid areas with cool, dry winters. Costs do not greatly exceed those of conventional aerial photography.

Of the various aerial photographic films, black and white infrared films sensitive in the near infrared spectrum often have slight advantages over conventional panchromatic and colour films, except that panchromatic infrared gives lighter crown tones in most vegetation, which reduce certain contrasts. Colour infrared (also known as false-colour, in which the invisible infrared radiation is assigned the colour red, while reflected red light is recorded as green, and reflected green light as blue) films may provide a greater measure of spectral discrimination. In particular, infrared films appear to enhance soil moisture and vegetation contrasts, and can be of real benefit in areas of low relief and dense cultivation where the evidence from landform or natural vegetation is weak or suppressed. As in the case of conventional panchromatic and colour films, the great advantages of photographic images are their high resolution, the facility for stereoscopic viewing and the relatively low cost. The best results are usually obtained from photography flown during the season when the water table is highest; this enhances the contrasts between soil textural classes and ensures that areas of seasonal seepage will be recorded.

In practice it is seldom possible to obtain widely different types of imagery from the same aerial mission, and the best single type must be chosen in relation to the terrain, scale of mapping and local climate. So, except for highly specialized applications, aerial photographs of one or other type are likely to be favoured over other techniques.

Plates 6 and 8–11 are annotated examples of various types of remote sensing imagery.

Appendix D. Geophysical techniques

Geophysical techniques for soil survey depend on there being measurable differences between the physical properties (e.g. magnetic susceptibility, electrical conductivity, density, elasticity, and thermal conductivity) of the soil materials to be distinguished. In some circumstances it may be possible to identify particular materials, but such interpretations are often ambiguous, so the surveyor usually matches geophysical records against field exposures or borings in order to interpret them and uses records mainly to interpolate the depth and thickness of significant layers between sample sites.

Static geophysical methods measure local distortions in static magnetic or gravitational fields of force. Dynamic methods involve the input of seismic and electromagnetic energy and the measurement of a response. The depth of penetration can be controlled by varying the spacing between energy source and the detectors to which the signals are returned. Relaxation methods, based on voltage decay in electrical signals, have not yet found significant application in engineering problems. Except for gravity surveys, all of these methods can be used in boreholes in order to calibrate the results obtained from intervening areas.

In every case the area is examined at sites on a preselected and staked grid; the grid interval depends on the complexity of the area and the purpose of the survey, but normally ranges from 20 to 200 m. The data from the observation points are usually presented as maps of iso-anomaly contours, or as geophysical profiles. Parasnis (1973) has discussed the various techniques in detail.

SEISMICS

Elastic or shock waves travel through rock or soil as longitudinal (P) waves and transverse (S) waves; different materials have characteristic transmission velocities for both of these. Seismic waves are generated by the shock of an explosion or a strong hammer blow on a steel plate in contact with the soil surface. A sparker is used to generate shock waves in water for a seismic survey beneath the bed of the sea or a river. As the waves pass downwards they are refracted at the boundaries between successive layers, and finally reflected back to be detected by electromagnetic geophones or other recorders at the surface. The time delay between the initial shock and the first return signals is related to the distance between source and recorder and used to calculate the thickness and velocity of transmission of various horizons in the rock or soil. The slope of the graph relating time to the distance travelled by the seismic wave gives the wave velocity of the material. Seismic velocities are highest in hard rock or strongly cemented materials and lowest in weathered or fractured rock and unconsolidated sediments. Closely spaced recorders detect shallow reflections and widely spaced recorders the reflections from deep layers. It may be necessary to use an array of geophones to match the wave patterns from different horizons. The signals are best recorded as wave profiles which show the initial shot and the 'kicks' produced at various geophones by the first wave returns from the refracting horizons, followed by the subsequent seismic traces (Fig. D.1). Other systems based only on the time to first wave returns (Fig. D.2), are regarded as open to error and misinterpretation.

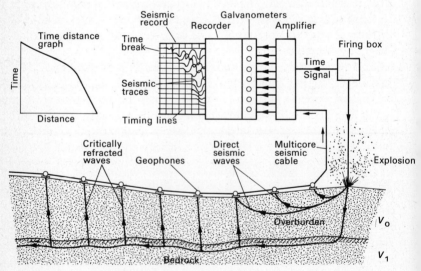

FIG. D.1. Principle of the multichannel seismic refraction method: two-layer case. (After Bullock 1978.)

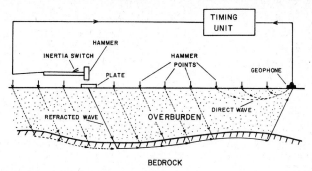

FIG. D.2. Principle of the hammer seismic method: two-layer case. (After Bullock 1978.)

The depth of the refractors and the wave transmission properties of intervening materials can be determined from these data as in Fig. D.3 which distinguishes soil and bedrock on their velocity. Although typical transmission velocities for various unconsolidated soil materials are known (Fig. D.4) it is better to check the identification of seismic horizons in field profiles, e.g. from trial holes or borings.

Modern multichannel seismic equipment normally uses 12 or 24 geophones spaced equidistantly in a straight line. The distance between the first and last geophones is known as the spread length. The spread length should be at least five times the desired depth of penetration (usually to bedrock). The signals recorded by the geophones may be enhanced to clarify the signal and to reduce extraneous 'noise'.

When a hammer blow is used to produce the energy the depth of penetration seldom exceeds 10 m (or perhaps 20 m when the record is enhanced). Penetration is smaller where dry, loose sands and gravels, or humic soils, attenuate the signal. However, depths in excess of 100 m can be achieved with explosives and modern, multichannel recording equipment.

Seismic surveys give best results when strongly contrasting materials are present and when the wave transmission rate increases with depth. Under normal circumstances the interface between two horizons cannot be identified clearly unless their transmission velocities differ by $100–200$ m s^{-1} corresponding to changes in bulk density of $0.2–0.4$ g cm^{-3}. Since soil horizons rarely show contrasts

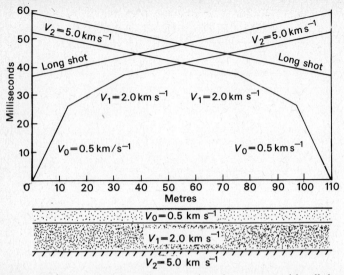

FIG. D.3. Seismic refraction diagram: three-layer case with all layers detected. (After Bullock 1978.) In seismic surveys, confirmation of the disposition of refracting horizons is obtained by measuring returns from different wave sources near either end of the spread (reverse shots). The left and right sloping graphs in this and subsequent diagrams (Figs. D.4–D.7) represent the recordings from these different sources. Long shots are achieved by detonating explosives some considerable distance from the ends of the spreads in order to achieve deeper penetration to bedrock in materials where rapid attenuation of the seismic waves occurs, e.g. through soils, and so to confirm the position of the soil/rock interface.

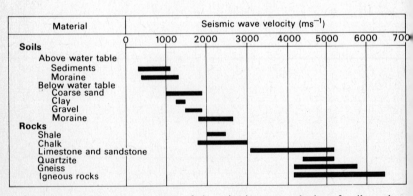

FIG. D.4. Approximate ranges of the seismic wave velocity of soils and rocks. (After West and Dumbleton 1975; modified from ABEM.)

of this magnitude, seismic surveys may show no more than the interface between soil and bedrock. Ambiguities may occur when the wave transmission rate decreases with depth, when contrasts are masked by saturation of the soil, or when the velocity of a layer changes laterally.

Estimates of depth to bedrock may be distorted where soft layers occur beneath harder materials, or very thin horizons are present, and the measured depth can be corrected only by calibration with recorded profiles. Layers thinner than about 2 m are not usually detected. The time/distance graphs which are produced from seismic field data can be interpreted quantitatively without undue difficulty (Hagedoorn 1959; Hawkins 1961). Their qualitative interpretation is more difficult: each break in the slope of the curve is equated with a deeper layer having a higher transmission velocity. Examples of time/distance graphs for different subsurface conditions are provided in Figs. D.5–D.7. The accuracy of seismic refraction surveys is generally about 15 per cent of the depth and is usually higher than that of any other geophysical method.

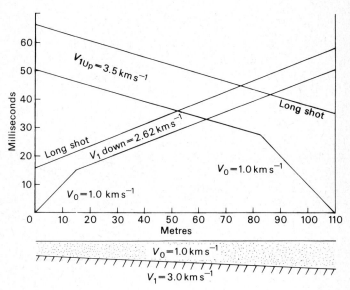

FIG. D.5. Seismic refraction diagram: two-layer case with dipping bedrock. (After Bullock 1978.)

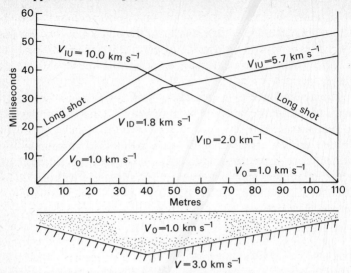

FIG. D.6. Seismic refraction diagram: two-layer case across buried valley. (After Bullock 1978.)

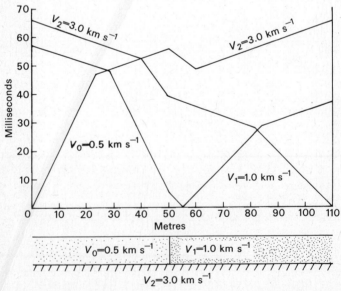

FIG. D.7. Seismic refraction diagram: two-layer case with change in soil velocity. (After Bullock 1978.)

The results of seismic surveys are usually presented as sections as in Fig. D.8, which shows the depths of interfaces between different materials and typical transmission velocities at several depths in each material. Given a few recorded profiles to calibrate these, it may be possible to interpolate the depth and thickness of un-consolidated material, or of key soil horizons between field obser-vation points. The technique has proved useful to determine the bedrock profile, especially the deep, buried channels beneath old river floodplains, and to predict the techniques and equipment most suitable for excavating different materials. Geological faults can often be identified, but cavities are rarely indicated.

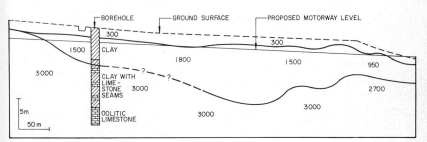

FIG. D.8. Section interpreted from a seismic survey. (After West and Dumbleton 1975.)

MAGNETICS

This technique is only applicable to materials containing significant amounts of the magnetically susceptible minerals magnetite, il-menite or pyrrhotite, and can only be used therefore to locate buried magnetic materials such as intrusions of igneous rocks and their weathering products.

Several small portable magnetometers are available. Most of these give instantaneous readings of magnetic flux in standard gamma units. Magnetic flux is noted at selected intervals along a series of traverse lines and plotted as a magnetic profile, e.g. Fig. D.9, from which the strike and extent of magnetic materials can often be identified. The dip of the magnetite body in Fig. D.9 is revealed by the asymmetry of the magnetic profile. Over intrusions

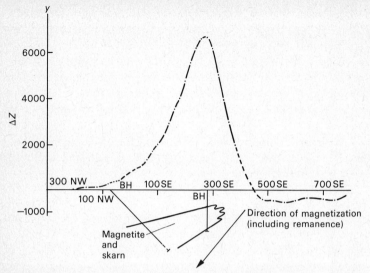

FIG. D.9. Profile across a magnetite body showing effect of transverse magnetization. (After Parasnis 1973, and by courtesy of E. Carlsson, Falun, Sweden.)

of basic igneous rocks the magnetic flux may be as high as 100 gammas, but unless background values are very uniform, identifications tend to be ambiguous at fluxes of less than approximately 25 gammas. If the anomaly is deep it is often difficult to locate the precise position of the edge of the magnetic material. Metal objects carried by the observer and pipes, fences and powerlines may interfere seriously.

The results of magnetometer surveys are usually presented as magnetic profiles. These may alert the surveyor to the presence of significant soil materials residual from igneous intrusions and the extent of such materials may be traced by a series of traverses.

GRAVITY

Variations in the gravitative attraction at the earth's surface are related to the densities of subsurface materials. The general value of the gravitational field at the earth's surface is approximately 980 gal (a gravitative acceleration of 1 cm s^{-2} is 1 gal). A sensitive gravimeter will measure small variations in subsurface density, such as are produced by variations in soil thickness, in milligal, but in most circumstances it is difficult to resolve gravity anomalies with an amplitude of less than 0.1 mgal (which approximates to a variation in soil thickness of some 4.5 m), unless they are of large extent, and then only if extraneous influences such as the effects of topography and variations in subsoil geology can be eliminated.

The direct reading at each observation point is first corrected for latitude, local elevation, terrain conditions and regional gravity trends. Even so the corrected or derivative (Bouguer) gravity anomalies must be calibrated against borehole data to interpret areas of fluctuating soil depth.

Gravity results are usually presented in the form of contour maps of Bouguer anomalies (e.g. Fig. 8.1, p. 180) or as contours of residual gravity after removing any regional trend which may obscure significant local variations. When satisfactorily calibrated such data can provide an accurate indication of fluctuations in soil depth but cannot distinguish between most soil horizons. Such information is particularly useful in karst areas, where the vertical distribution of soil and rock materials in relation to the level of the water table is used to assess the probability of sinkhole and subsidence formation, but the procedure rarely identifies cavities unless they are very large or very shallow.

RESISTIVITY

This technique involves the introduction of a current into the soil; a sensitive voltmeter measures the potential difference between a pair of electrodes in the soil, from which can be calculated values of resistivity measured in ohm centimetres. Resistivity is a function of the electrolyte in the pore spaces of a soil, and so is inversely proportional to its porosity. The depth at which the potential is measured is proportional to the spacing of the current electrodes, provided that the geoelectrical conditions are homogeneous, which is seldom the case. A resistivity profile may be recorded at each observation point (taken to be the midpoint between the potential electrodes) by making a set of measurements with increasing electrode separation. The resistivity measured at each electrode spacing is not the true resistivity, but an apparent value which depends not only on the properties of the material but also on the thickness of the layer being measured. The electrodes are most commonly equidistant with the current electrodes outside the potential electrodes (the Wenner array), but other arrays such as the Schlumberger are sometimes used (Fig. D.10). The Schlumberger array gives more accurate results and deeper penetration with the same electrode spacing than does the Wenner array. The spread length, or distance between extreme electrodes, is not usually less than five times the required depth of penetration.

The results of each resistivity probe are plotted as a sounding curve with apparent resistivity as ordinate and electrode separation (as a measure of depth) as abscissa; breaks in the curve usually

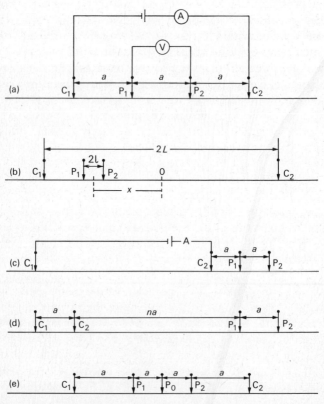

FIG. D.10. Some electrode configurations commonly used in resistivity surveys. (a) Wenner. (b) Schlumberger. (c) Three-point. (d) Dipole–dipole. (e) Lee. (After Parasnis 1973.)

indicate changes in subsurface material or in moisture content. Another method is to maintain a constant electrode separation to provide information at a specific depth, and to move the entire electrode array from station to station in order to identify lateral changes in resistivity. It is not usually possible to identify soil materials from their resistivity, but it is possible to identify them against materials observed in exposures or trial holes. Soil resistivity ranges typically from 10^1 to nearly 10^7 Ω cm; some characteristic values for different soils are given in Fig. D.11. The record may be distorted or obliterated by the presence of water in the soil.

Sounding curves are best interpreted by (i) determining the number of layers present by comparing several adjacent curves, and (ii) calibrating these against theoretically calculated curves to obtain local resistance and conductivity values. This is a skilled procedure and it requires considerable experience to avoid ambiguous interpretations. The depth to boundaries between layers can seldom be determined to an accuracy better than 20 per cent.

The results of resistivity surveys are usually presented as sections

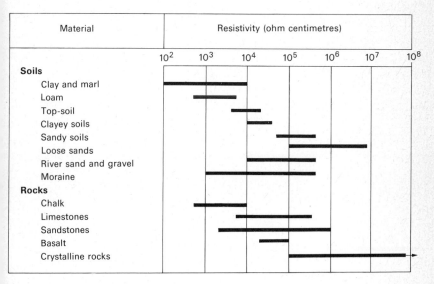

FIG. D.11. Approximate ranges of resistivity of soils and rocks. (After West and Dumbleton 1975; modified from ABEM.)

showing resistivity values and interpreted interfaces between different soil types (Fig. D.12). These are particularly useful for interpolating soil discontinuities between field observation points, since differences between horizons are often resolved better by this technique than by seismic methods.

ELECTROMAGNETICS

An alternating electromagnetic field induces currents in conducting materials in its path and may be used to identify faults, fractures, and weathered dykes which contain conductive water.

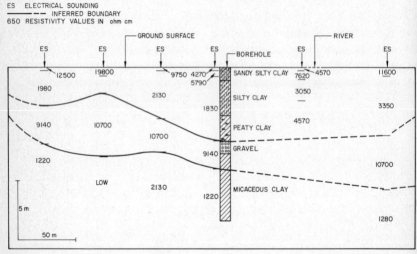

FIG. D.12. Section interpreted from a resistivity survey. (After West and Dumbleton 1975.)

Appendix E. Description of rock masses for engineering

Although this book specifically deals with soil survey procedures, the description and analysis of rock masses is an integral part of engineering geology and geotechnical engineering. As was pointed out in Chapter 2, the division between soil and rock, which we have equated with an unconfined compressive strength of 700 kPa, is an arbitrary value within a continuum of increasing consistency or hardness. However, the mechanical behaviour of rock masses has been found to differ in many respects from that of soils, particularly in the tendency of most rocks to behave in a brittle manner under the relatively low pressures to which they are subject in shallow engineering works. The analysis of the behaviour of rock masses falls within the field of rock mechanics, and will not be considered further here. However, since bedrock is frequently encountered in the course of engineering soil surveys, it is important that the observer should be able to describe rock samples or exposures systematically so that valuable information is not lost. Moreover, such description should not be restricted to individual intact rock fragments, but should be applied to the entire fabric of the rock mass including discontinuities such as joints and other fractures, which often exert a significant influence on the strength of the mass and its engineering behaviour.

Undisturbed rock encountered during soil surveys is usually restricted to natural exposures; some disturbance of the rock fabric is generally present in excavated exposures such as test pits and trial holes and in the core samples recovered in the course of rotary core drilling (pp. 281–7). The rock chips or cuttings produced by percussion drilling are too small to give any worthwhile information on discontinuities or fractures in the rock mass. The extent of the disturbance must obviously be taken into account when interpreting the engineering properties of the rock mass; for example, excavated exposures and core samples often contain artificial fractures produced by the high forces needed to penetrate the rock.

The system proposed below for the description of rock masses is compatible with the methods of soil profile description described in Chapter 6. It is based on qualitative description and simple manual tests in the field in order that it can be easily and widely applied under a variety of conditions. Its aim is to enable the reader to visualize the rock mass and to help him predict its behaviour in excavations or under applied loads. Several other systems have been proposed, e.g. the Geological Society Engineering Group Working Party (1977) and the Core Logging Committee (1978); the present system is closely akin to the latter.

A clear distinction is made at the outset between descriptions of the rock mass *in situ* and descriptions of core samples. With the former, variations in the rock fabric and the disposition of fractures are usually recorded by means of a 'line survey' in which the appropriate parameters are related to distances along a pre-selected line in a trench or across an outcrop area; core samples, by their nature, usually conform to a vertical or steeply inclined line from the ground surface through the rock mass. Also, unless laborious and expensive procedures are used to recover 'orientated' cores, it is not usually possible to describe the actual directions in the rock mass of bedding and discontinuities in the rock fabric, and the lengths over which most discontinuities persist cannot be determined from the limited sample provided by a drilling core. However, useful information may be obtained from the records kept during drilling; for example, the rate of penetration can be a guide to the hardness of the various strata, core losses indicate the presence of voids or soil materials which are not easily recovered in the drilling process, and losses of drilling fluid reflect the existence of openings or permeable materials. A separate *borehole log* should be kept to record these data as well as information on the drilling procedure and drill diameter, water table level, the depths at which samples were recovered or *in situ* tests carried out, and the chief results obtained from any tests. This borehole log should not be confused with the *core log*, in which the various characteristics of the rock samples are factually described. Any inferences made by the describer, for example interpretations of the likely nature of soil materials represented by core losses, should be clearly distinguished from such factual data.

The equipment required by the system proposed is limited to a measuring tape, a pocket knife, and a geological hammer; more sophisticated devices such as a Schmidt rebound hammer or a portable jack for point-load strength testing may be used to make a more accurate assessment of rock hardness in the higher ranges (Geological Society Engineering Group Working Party 1977). The rock should, whenever possible, be described from wetted, freshly broken surfaces in order that the properties of the freshest material are clearly revealed.

The full description of the rock mass is divided into three parts:

(i) description of the body of the rock;
(ii) description of the fracture surfaces;
(iii) description of the fracture fillings.

DESCRIPTION OF THE BODY OF THE ROCK

Colour

Colour is a useful guide to rock composition and state of weathering. Munsell colour charts are preferred to qualitative assessments. As with soils, the colouring is often complex and variations may be introduced by banding, veining, staining of fracture faces, mottling, etc.

Weathering

The following five classes are advocated in conformity with the proposals of the Core Logging Committee (1978):

Unweathered

No visible signs of alteration in the rock material, but fracture planes may be stained or discoloured.

Slightly weathered

Fractures are stained or discoloured and may contain a thin filling of altered or illuviated material. Discoloration may extend into the rock from the fracture planes to a distance of up to 20 per cent of the fracture spacing (i.e. less than 40 per cent of the core is discoloured).

Moderately weathered

Slight discoloration extends from the fracture planes for a distance greater than 20 per cent of the fracture spacing (i.e. through the greater part of the rock). Fractures may contain filling of altered or illuviated material. The surface of the rock is not friable (except in the case of poorly cemented sedimentary rocks) and the original texture of the rock is preserved. Grain boundaries are sometimes partially open.

Highly weathered

Discoloration extends throughout the rock, and the surface is friable. In rock cores the surface is usually pitted due to the washing out of highly altered minerals by the drilling fluid. The original texture of the rock has generally been preserved but separation of grains has occurred.

Completely weathered

The rock is totally discoloured and the external appearance of excavated blocks or lengths of core is that of a residual soil. Internally the rock texture is partly preserved but grains have separated completely. Such residual material should be described in terms of the normal soil descriptors MCCSSO (Chapter 6).

Texture

In most soils aggregates of particles can be broken down between the fingers without difficulty and a tactile assessment of the dominant particle sizes can be made; this is not possible with rock and a field description must be based on a visual assessment. Experience indicates that this kind of appraisal tends to bias the average texture in favour of the coarser particle sizes. The following descriptors and class intervals have been selected with this bias in mind so as to conform in a general way with the principal classes used in the engineering description of soils:

> Very coarse: grains or crystals >6.0 mm
> Coarse: grains or crystals $2.0–6.0$ mm
> Medium: grains or crystals $0.6–2.0$ mm
> Fine: grains or crystals $0.2–0.6$ mm
> Very fine: grains or crystals <0.2 mm

Microstructure and fracture spacing

Bedding plane spacing

Under this heading are also included foliation in metamorphic rocks and flow banding in igneous rocks:

Very thickly bedded >2.0 m
Thickly bedded 0.6–2.0 m
Medium bedded 0.2–0.6 m
Thinly bedded 60 mm–0.2 m
Very thinly bedded 20–60 mm
Laminated 6–20 mm
Thinly laminated <6 mm

For the purposes of this system of description fractures include all mechanical discontinuities such as joints, faults, shear-zones, bedding planes, foliation, and cleavage planes and flow bands, if these represent planes of weakness. The spacing is measured normal to the inclination of the fractures or the various fracture sets.

Fracture spacing

		Fractures per m
Very slightly fractured	>2.0 m	<1
Slightly fractured	0.6–2.0 m	about 1
Moderately fractured	0.2–0.6 m	1–5
Highly fractured	20 mm–0.2 m	5–50
Very highly fractured	<20 mm	>50

Hardness

The various categories of rock hardness are identified as follows:

Very soft rock

Material crumbles under firm blows with the sharp end of geological pick and can be peeled with a knife; too hard to cut an undisturbed sample by hand. Typical unconfined compressive strength (UCS): 700–4000 kPa.

Soft rock

Can just be scraped and peeled with a knife; indentations 1–3 mm deep are formed in the specimen by firm blows of the pick point. Typical UCS: 4000–10 000 kPa.

Hard rock

Cannot be scraped or peeled with a knife. Hand-held specimen can be broken with a single firm blow of the hammer end of a geological pick. Typical UCS: 10 000–20 000 kPa.

Very hard rock

Hand-held specimen breaks after more than one blow of the hammer end of a pick. Typical UCS: 20 000–70 000 kPa.

Extremely hard rock

Many blows with geological pick required to break specimen. Typical UCS: > 70 000 kPa.

Rock type

The rock type should be identified in terms of the criteria given for various categories of igneous, metamorphic, and sedimentary rocks in Chapter 3. Like the origin of a soil, the stratigraphic horizon to which a rock belongs often provides a useful guide to the engineering properties of the rock mass, and should be identified wherever possible.

DESCRIPTION OF THE FRACTURE SURFACES

Fracture type

The main fracture types are listed above. The identification of the origin of the fracture is important in assessing its engineering significance.

Separation of faces

This has a marked effect on the strength of the rock mass:

Closed	0 mm
Very narrow	0–0.6 mm
Narrow	0.6–2.0 mm
Wide	2.0–6.0 mm
Very wide	6.0–20 mm
Major fracture	> 20 mm

Staining or filling

These are identified as follows:

Clean: no filling material
Stained: coloration of fracture faces only. No recognizable filling material.
Filled: recognizable filling material.

Description of surfaces

The classification proposed by the Core Logging Committee (1978) is advocated:

Smooth: appears smooth and is generally smooth to the touch. May be slickensided (record direction).
Slightly rough: asperities on the fracture surfaces are visible and can be distinctly felt.
Medium rough: asperities are clearly visible and fracture surface feels abrasive.
Rough: large angular asperities are visible, some in the form of ridges and steps with high side angles.
Very rough: near vertical steps and ridges occur on the fracture surface.

Orientation

In unorientated cores the relative dip of the various fracture sets in relation to the horizontal is recorded. In rock masses the absolute dip and direction of dip of the various sets are tabulated.

Persistence

This can only be described in well-exposed rock masses and is difficult to quantify. The most meaningful value is the average trace length of fractures belonging to various sets.

Form of intervening blocks

When the rock mass can be examined in three dimensions, the forms outlined by fractures can be classified. The following categories are proposed by the Engineering Group Working Party (1977):

Blocky: approximately equidimensional;
Tabular: one dimension considerably shorter than the other two;
Columnar: one dimension considerably longer than the other two.

DESCRIPTION OF THE FRACTURE FILLING

If the fractures are filled, details of the filling should, where possible, be appended at the end of the description using the same descriptors that are used to record the soil profile (Chapter 6); fracture fillings often influence the strength of the rock mass, and the extent of this influence is often predictable from a proper description. Identification of the *type* of filling is often useful, e.g. iron oxides, calcite, chlorite, silt or clay, siliceous material, etc., as is its *origin*, e.g. fault gouge, cemented vein filling, illuviated infilling, or residual coating.

Borehole and core logs

In addition to the above descriptors, a core log prepared from borehole samples usually includes several additional parameters:

Core recovery

The percentage of each core run (i.e. the distance over which the borehole was deepened between withdrawing the core-barrel and extracting the core) which is represented by solid material. Where losses of core are recorded, their precise position should, whenever possible, be indicated on the core log.

Rock quality designation (RQD)

This is the percentage of the length of core recovered in a drill-run which is in the form of pieces of fully circular core, each with a length of more than 100 mm. Some important rock mechanics indices can be correlated with the RQD value.

Fracture frequency

This is the number of fractures which intersect a unit length of rock (either a length of a core or of a particular axis selected to traverse the rock mass). Typical ranges of fracture frequency have been given on p. 353.

Examples of borehole and core logs prepared in accordance with the above system are given in Fig. E.1.

It should be noted that descriptions of this type form the basis of several rock classification systems for engineering (e.g. Barton, Lien, and Lunde 1974; Bieniawski 1973; Cottiss, Dowell, and Franklin 1971; Deere, Merritt, and Coon 1969) by means of which the suitability of a particular rock mass for various engineering works can be assessed as an aid to site selection and engineering design.

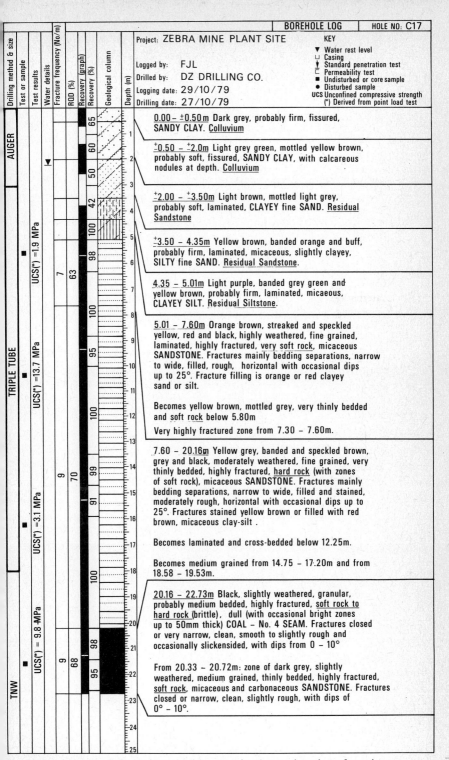

FIG. E.1. Borehole log prepared by examination and testing of continuous core samples obtained by rotary drilling.

References

Abbiss, C. (1979). Sounding out the ground. *BRE News* 44, Summer 1978. Building Research Establishment, Garston, England.

Acker, W. L. (1974). *Basic procedures for soil sampling and core drilling.* Acker Drill Co. Inc., Scranton, USA.

Aitchison, G. D., and Grant, K. (1967). The PUCE programme of terrain description, evaluation and interpretation for engineering purposes. *Proc. 4th Reg. Conf. Africa Soil Mechanics and Foundation Engineering,* Cape Town.

Allen, H. (1945). Report of Committee on Classification of Materials for Subgrades and Granular Type Roads. *Proc. Highway Res. Bd,* Washington, **25**, 375–92.

American Society for Testing and Materials (1974). *1974 Annual book of ASTM Standards.* ASTM, Philadelphia, USA.

Anderson, R. V. (1973). Residential subdivision services. *Proc. natn Conf. Urban Engineering Terrain Problems.* Tech. Memo. 109, pp. 189–201. National Research Council of Canada, Ottawa.

Anuta, P. E., Kristof, S. J., Levandowski, D. W., Phillips, T. L., and MacDonald, R. B. (1971). Crop, soil and geological mapping from digitized multispectral satellite photography. *Proc. 7th Int. Symp. Remote Sensing of the Environment,* Ann Arbor, Michigan, 1983–2016.

Atkinson, J. H., and Bransby, P. L. (1978). *The mechanics of soils. An introduction to critical state soil mechanics.* McGraw-Hill, London.

Atterberg, A. (1911). Die Plastizitat der Tone. *Int. Mitt. Bodenk.* **1**, 10–43.

—— (1912). Die Plastizitat der Tone. *Int. Mitt. Bodenk.* **2**, 149–89.

Attewell, P. B., and Farmer, I. W. (1976). *Principles of engineering geology.* Chapman and Hall, London.

Barton, N., Lien, R., and Lunde, J. (1974). Engineering classification of rock masses for the design of tunnel support. *Rock Mech.* **6**, 189–236.

Beaumont, T. E. (1977). Techniques for the interpretation of remote sensing imagery for highway engineering purposes. Transport and Road Research Laboratory Report 753. Department of the Environment, Department of Transport, Crowthorne, Berkshire, England.

Beckett, P. H. T. (1971). Output from a terrain data store. Report 71506. Military Vehicles and Engineering Establishment, Christchurch, England.

—— Webster, R., McNeil, G. M., and Mitchell, C. W. (1972). Terrain evaluation by means of a data bank. *Geog. J.* **138**, 430–56.

Bieniawski, Z. T. (1973). Engineering classification of jointed rock masses. *Trans. S. Afr. Instn. civ. Engrs* **15**, 335–343

Bishop, A. W., Alpan, I., Blight, G. E., and Donald, I. B. (1960). Factors controlling the strength of partly saturated soils. *Proc. Res. Conf. Shear Strength of Cohesive Soils*, Am. Soc. civ. Engrs, Colorado, pp. 503–32.

Bjerrum, L. (1955). Stability of natural slopes in quick clay. *Géotechnique* **5**, 101–19.

Blight, G. E. (1970). *In situ* strength of rolled and hydraulic fill. *J. Soil Mech. Fdns Div. Am. Soc. civ. Engrs*, 881–99.

Bourne, R. (1931). Regional survey and its relation to stocktaking of the agricultural resources of the British Empire. *Oxf. For. Mem.* **13**, 16–18.

Boussinesq, V. J. (1885). *Application des potentiels a l'étude de l'equilibre, et du mouvement des solides elastiques avec des notes etendues sur divers points de physique mathematique et d'analyse.* Gauthier-Villars, Paris.

Brackley, I. J. A. (1979). Prediction of heave under foundations. Ground Profile. *Newsl. Div. geotech. Engng S. Afr. Instn civ. Engrs*, **19**, 20–3

Brink, A. B. A. (1979). *Engineering geology of Southern Africa*. Building Publications, Silverton.

——, and Partridge, T. C. (1965). Transvaal karst: some considerations of development and morphology with special reference to sinkholes and subsidences in the Far West Rand. *S. Afr. geogr. J.* **67**, 11–33.

——, Mabbutt, J. A., Webster, R., and Beckett, P. H. T. (1966). Report of the working group on land classification and data storage. Report 940, Military Engineering Experimental Establishment, Christchurch, England.

——, Partridge, T. C., Webster, R., and Williams, A. A. B. (1968). Land classification and data storage for the engineering usage of natural materials. *Proc. 4th Conf. Aust. Road Res. Bd*, Vol. 4, Part 2, pp. 1624–47.

British Standards Institution (1957). Site investigations. British Standard Code of Practice CP 2001. London.

—— (1975). Methods of test for soils for civil engineering purposes. British Standard, 1377. London.

Brown, R. J. E. (1970). *Permafrost in Canada*. University of Toronto Press.

——, and Johnston, G. H. (1964). Permafrost and related engineering problems. Technical paper 173. Division of Building Research, National Research Council of Canada.

Building Research Station (1968). *Concrete in sulphate-bearing soils and groundwaters*. Digest 90. HMSO, London.

Bullock, S. J. (1978). The case for using multi-channel seismic equipment and techniques for site investigation. *Bull. Ass. engng Geolog.* **XV**, 19–35.

Burland, J. B. (1979). Discussion. Application of geotechnics to the solution of engineering problems—essential preliminary steps to relate the structure to the soil which provides its support. *Proc. Instn civ. Engrs* **66**, 497–8.

——, and Wroth, C. P. (1974). Settlement of buildings and associated damage. Settlement of structures. *Proc. Conf. Br. Geotech. Soc.*, Cambridge, pp. 611–54. Pentech Press, London.

——, Broms, B., and de Mello, V. F. B. (1977). Behaviour of foundations and structures. State-of-the-art report. *Proc. 9th Conf. Soil Mechanics and Foundation Engineering*, Tokyo, Vol. 2, pp. 495–546.

Casagrande, A. (1947). Classification and identification of soils. *Proc. Am. Soc. civ. Engrs* **73**, 783–810.

Christian, C. S., and Stewart, G. A. (1953). General report on survey of Katherine–Darwin region, 1946. CSIRO Australian Land Resources Series 1.

Corcoran, P., Jarvis, M. G., Mackney, D., and Stevens, K. W. (1977). Soil corrosiveness in South Oxfordshire. *J. Soil Sci.* **28**, 473–84.

Core Logging Committee (1978). A guide to core logging for rock engineering. *Bull. Ass. engng Geolog.* **15**, 295–328.

Cottiss, G. I., Dowell, R. W., and Franklin, J. A. (1971). A rock classification system applied to civil engineering. *Civ. Engng publ. Wks Rev.* **66**, Part 1, 611–14; Part 2, 736–43.

De Beer, J. H. and Biggs, D. C. (1978). Urban geotechnical data banking. *Proc. 3rd Int. Cong. Int. Ass. engng Geolog.*, Madrid. Special Session 4, pp. 130–7.

Deere, D. V., Merritt, A. H., and Coon, R. F. (1969). Engineering classification of *in situ* rock. Technical Report AFWL-67-144. United States Air Force System Command, Kirkland Air Force Base, New Mexico.

Dowling, J. W. P. and Beaven, P. J. (1969). Terrain evaluation for road engineers in developing countries. *J. Instn Highw. Engrs* 5–15.

Frankipile (1979). Franki facts No. 11. Maharani Hotel, Durban. Frankipile South Africa, Johannesburg.

Fugro, N. V. (1974). In Comparisons of the results from static penetration tests and large *in situ* plate tests in London Clay (ed. A. Marsland). *Proc. Eur. Symp. Penetration Testing*, Stockholm. [Current paper 87/74. Building Research Establishment, England.]

Geological Society Engineering Group Working Party (1977). The description of rock masses for engineering purposes. *Q. Jl engng Geol.* **10**, 355–88.

Glossop, R. (1968). The rise of geotechnology and its influence on engineering practice. 8th Rankine Lecture. *Géotechnique* **18**, 105–50.

Grant, K., and Lodwick, G. D. (1968). Storage and retrieval of information in a terrain classification system. *Proc. 4th Conf. Aust. Road Res. Bd*, **4**, Part 2.

Hagedoorn, J. G. (1959). The plus–minus method of interpreting seismic refraction sections. *Geophys. Prospect.* **26**, 158–82.

Hawkins, L. H. (1961). The reciprocal method of routine seismic refraction investigations. *Geophysics* **26**, 806–19.

Helmer, R. A. (1959). *Soils manual of the State of Oklahoma Department of Highways*, p. 37.

Hodgson, J. M. (1978). *Soil sampling and soil description*. Clarendon Press, Oxford.

Hughes, J. M. O., Wroth, C. P., and Windle, D. (1977). Pressuremeter tests in sands. *Géotechnique* **27**, 455–77.

Hvorslev, M. J. (1949). Subsurface exploration and sampling of soils for civil engineering purposes. US Army Corps of Engineers, Waterways Experiment Station, Vicksburg, Mississippi.

Jennings, J. E. (1966). Building on dolomites in the Transvaal. *Trans. S. Afr. Instn. civ. Engrs* **8**, 41–62.

——, and Brink, A. B. A. (1978). Application of geotechnics to the solution of engineering problems—essential preliminary steps to relate the structure to the soil which provides its support. *Proc. Instn civ. Engrs* **64**, 571–89.

——, and Evans, G. A. (1962). Practical procedures for building in expansive soil areas. *S. Afr. Bldr.*

——, and Knight, K. (1956). Recent experiences with the consolidation test as a means of identifying conditions of heaving or collapse of foundations on partially saturated soils. *Trans. S. Afr. Instn civ. Engrs* **6**, 255–6.

——, —— (1957). The additional settlement of foundations due to collapse of structure of sandy subsoils on wetting. *Proc. 4th Int. Conf. Soil Mechanics and Foundation Engineering*, London, Vol. 1, pp. 316–19.

——, —— (1975). A guide to constructing on or with materials exhibiting additional settlement due to collapse of grain structure. *Proc. 6th Reg. Conf. Afr. Soil Mechanics and Foundation Engineering*, Durban, Vol. 1, pp. 99–105.

——, Brink, A. B. A., and Williams, A. A. B. (1973). Revised guide to soil profiling for civil engineering purposes in Southern Africa. *Trans. S. Afr. Instn civ. Engrs* **15**, 3–12.

Jones, G. A. (1978). Cone penetration test. Lecture 4. Course on *in situ* testing in boreholes. South African Institution of Civil Engineers and National Institute of Transport and Road Research, Pretoria.

Kantey, B. A., and Williams, A. A. B. (1962). The use of soil engineering maps for road projects. *Trans. S. Afr. Instn civ. Engrs* **4**, 149–59.

King, L. C. (1972). *The Natal monocline: explaining the origin and scenery of Natal, South Africa*. Department of Geology, University of Natal, Durban.

Komornik, A. (1979). Personal communication.

Lawrance, C. J. (1972). Terrain evaluation in West Malaysia. Part 1. Terrain classification and survey methods. Report LR506, Transport and Road Research Laboratory, Department of the Environment, Britain.

—— (1975). The use of punched cards in the storage and retrieval of engineering information. Ethiopian Highway Authority and Transport and Road Research Laboratory UK. Joint Road Research Project, Report 2.

——, Webster, R., Beckett, P. H. T., Bibby, J. S., and Hudson, G. (1977). The use of airphoto interpretation for land evaluation in the Western Highlands of Scotland. *Catena* **4**, 341–57.

Leggett, R. F. (1973). *Cities and geology*. McGraw-Hill, New York.

—— (1979). Geology and geotechnical engineering. 13th Terzaghi lecture. *J. Geotech. Engng Div.* **105** (GT3), 339–91.

362   References

Leonards, G. A. (Ed.) (1962). *Foundation engineering*. McGraw-Hill, New York.

Linell, K. A., and Tedrow, J. C. F. (1981). *Soil and permafrost surveys in the Arctic and subarctic*. Clarendon Press, Oxford.

Linton, D. L. (1951). The delimitation of morphological regions. In *London essays in geography*. Longmans Green, London.

Lumb, P. (1965). The residual soils of Hong Kong. *Géotechnique* **15**, 180–94.

Miller, J. P., and Scholten, R. (1962). *Sedimentary rocks*. Laboratory studies in geology 215. Freeman, San Francisco.

Mitchell, C. W., Webster, R., Beckett, P. H. T., and Clifford, B. (1979). An analysis of terrain classification for long-range prediction of conditions in deserts. *Geogrl J.* **145**, 72–85.

Morse, R. K. (1961). *Engineering properties of the superficial soils of Livingstone County, Illinois*. University of Illinois, Urbana.

——, and Thornburn, T. H. (1961). Reliability of soil map units. *Proc. 5th Int. Conf. Soil Mechanics and Foundation Engineering*, Paris, Vol. 1, pp. 259–62.

Netterberg, F. (1971). Calcrete in road construction. CSIR Research Report 286. National Institute of Road Research, Pretoria.

——, and Maton, L. J. (1975). Soluble salt and pH determinations on highway materials. *Proc. 6th Reg. Conf. Afr. Soil Mechanics and Foundation Engineering*, Durban, 1, pp. 131–9.

——, ——, and De Kock, T. J. (1975). Estimation of sulphate content of crusher runs from total salt and pH measurements. *Proc. 6th Reg. Conf. Afr. Soil Mechanics and Foundation Engineering*, Durban, Vol. 1, pp. 141–3.

Oberholster, R. E., and Brandt, M. P. (1975). Transmission and scanning electron micrographs of some selected phyllosilicates and inosilicates of southern Africa. CSIR Research Report BOU 32. Pretoria.

Palmer, D. J. (1957). *Writing reports*, pp. 1–24. Soil Mechanics Ltd., London.

Parasnis, D. S. (1973). *Mining geophysics*, 2nd edn. Elsevier, Amsterdam.

Parry, R. H. G. (1978). Stability of clay linings. *Proc. Instn civ. Engrs* **65**, 271–82.

Partridge, T. C. (1975). Some geomorphic factors influencing formation and engineering properties of soil materials in South Africa. *Proc. 6th Reg. Conf. Africa Soil Mechanics and Foundation Engineering*, Durban, Vol. 1, pp. 37–42.

——, Brink, A. B. A., and Mallows, E. W. N. (1973). Morphological classification and mapping as a basis for development planning. *S. Afr. geogr. J.* **55**, 69–80.

Pike, C. W., and Saurin, B. F. (1952). Bouyant foundations in soft clay for oil refinery structures at Grangemouth. *Proc. Instn civ. Engrs* **1**, 301–34.

Rauch, H. P. (1963). Soil mechanics applied to railway earthworks. *Proc. Diamond Jubilee Conv. S. Afr. Instn civ. Engrs*, pp. 155–61.

Road Research Laboratory (1961). *Soil mechanics for road engineers.* HMSO, London.

Roscoe, K. H., Schofield, A. N., and Wroth, C. P. (1958). On the yielding of soils. *Géotechnique* **8**, 22–53.

Rowe, P. W. (1962). The stress dilatancy relation for static equilibrium of an assembly of particles in contact. *Proc. R. Soc.* **A269**, 500–27.

—— (1971). Theoretical meaning and observed values of deformation parameters for soils. *Proc. Roscoe Mem. Symp.*, pp. 143–94.

Sanglerat, G. (1972). *The penetrometer and soil exploration.* Elsevier, Amsterdam.

Schmertmann, J. H. (1970). Static cone to compute static settlement over sand. *J. Soil Mech. Fdns Div. Am. Soc. civ. Engrs* **96**, 1011–43.

Schofield, R. K. (1935). The pF of the water in soil. *Trans. 3rd Int. Congr. Soil Science*, Oxford, Vol. 2, pp. 37–48.

Sherard, J. L., and Decker, R. S. (Eds.) (1977). Dispersive clays, related piping, and erosion in geotechnical projects. ASTM Special Technical Publication 623. American Society for Testing and Materials.

Sherard, J. L., Woodward, R. J., Giziensky, S. F., and Clevenger, W. A. (1963). *Earth and earth–rock dams.* Wiley, London.

Skempton, A. W. (1949). Vane tests in the alluvium plain of the River Forth near Grangemouth. *Géotechnique* **1**, 111–24.

——, and Hutchinson, J. N. (1969). Stability of natural slopes and embankment foundations. *Proc. 7th Int. Conf. Soil Mechanics and Foundation Engineering*, Mexico, Vol. **3**, p. 381.

Smith, R. T., and Atkinson, K. (1975). *Techniques in pedology: a handbook for environmental and resource studies.* Elek, London.

South African Institution of Civil Engineers (1980). Code of practice relating to the safety of men working in small diameter vertical and near vertical shafts for civil engineering purposes. *Trans. S. Afr. Instn civ. Engrs* in press.

Terracina, F. (1962). Foundations of the tower of Pisa. *Géotechnique* **12**, p. 336–9.

Terzaghi, K. (1936). The shearing resistance of saturated soils. *Proc. 1st Int. Conf. Soil Mechanics*, Cambridge, Mass., Vol. 1, pp. 54–6.

—— (1955). Influence of geological factors on the engineering properties of sediments. *Econ. Geol.* **50**, 557.

——, and Peck, R. B. (1967). *Soil mechanics in engineering practice.* Wiley, New York.

Tomlinson, M. J. (1963). *Foundation design and construction.* Pitman, London.

Tomlinson, R. F. (1968). A geographic information system for regional planning. In *Land evaluation.* CSIRO Symp., Canberra. Macmillan, Melbourne.

United States Army Corps of Engineers (1953). The Unified Soil Classification System. Technical Memorandum 3.357, Vols. 1 and 3.

United States Bureau of Reclamation (1960a). Design of small dams. Government Printing Office, Washington.

—— (1960b). Earth manual. Government Printing Office, Washington.

Unstead, J. F. (1933). A system of regional geography. *Geography* **18**, 185–7.

Van Der Merwe, D. H. (1964). The prediction of heave from the Plasticity Index and percentage clay fraction. *Trans. S. Afr. Instn civ. Engrs* **6**, 103–7.

——, and Savage, P. F. (1979). Nomogram for the prediction of heave. Personal communication.

Van Vleck, E. M., Sinclair, K. F., Pitts, I. W., and Slye, R. E. (1973). Earth resources ground data handling systems for the 1980s. NASA Tech. Memo. NASA TMX-62, 240.

Webster, R. (1977). *Quantitative and numerical methods in soil classification and survey*. Clarendon Press, Oxford.

Weinert, H. H., and Dehlen, G. L. (1965). Appraisal of road building materials. *Proc. 2nd Symp. Soil Eng. Mapping and Data Storage*, SAICE, Thatchstone Inn. Nat. Building Research Inst., CSIR, Pretoria.

West, G., and Dumbleton, M. J. (1975). An assessment of geophysics in site investigations for roads in Britain. Transportation and Road Research Laboratory Report LR680.

White, L. P. (1977). *Aerial photography and remote sensing for soil survey*. Clarendon Press, Oxford.

Wilkinson, W. B. (1968). Constant head *in situ* permeability tests in clay strata. *Géotechnique* **18**, 172–94.

Willbourne, J. (1972). Foundations for the London Central YMCA. *Civ. Engng publ. Wks Rev.* 257–8.

Windle, D., and Wroth, C. P. (1977). The use of a self-boring pressure-meter to determine the undrained properties of clays. *Ground Engng* 37–46.

Winterkorn, H. F., and Fang, H-Y. (1975). *Foundation engineering handbook*. Van Nostrand Reinhold, New York.

Glossary of symbols

In a few cases a symbol is used to represent two or more different quantities. In general these different quantities do not appear in the same section so that confusion should not occur if the context is considered.

a Radius.

c Cohesion, cohesive strength.

c' Cohesion in terms of effective stress.

c'_r Residual cohesion.

c_u Cohesion under undrained conditions.

c_v Coefficient of consolidation.

C_c Compression index.

C_p Velocity of P-waves.

C_s Swelling index: velocity of S-waves.

d Length of drainage path.

D diameter.

e Void ratio.

e_0 Initial void ratio.

e_1, e_2 Values of void ratio.

E Young's modulus.

E' Equivalent modulus of compressibility

E_u Undrained elastic modulus (Young's modulus).

f A reduction factor; slope of a curve.

F Factor of safety: intake factor.

G Shear modulus; specific gravity of particles.

h Head (or difference in height).

H Height of dimension; thickness of layer.

I_p The plasticity index.

k D'Arcy coefficient of permeability; modulus of sub-grade reaction.

K Bulk modulus; a pore-pressure factor; dilatancy factor.

K_0 Earth pressure coefficient.

l Length.

L Length dimension.

LL   Liquid limit.

LS	Linear shrinkage.
m_v	Coefficient of compressibility.
M_d	Disturbing moment.
N	The number of blows per foot in a standard penetration test (SPT).
n	A bearing capacity factor
p	Mean stress.
p_c	Pre-consolidation pressure.
p	Initial pressure.
p_1, p_2	Values of pressure.
p'	Mean effective pressure.
p'_1, p'_2	Effective pressures.
PI	Plasticity index.
PL	Plastic limit.
q	Deviator stress; applied pressure.
q'	Deviator stress.
$q_c, q_{c0},$	Resistances to cone penetration.
q_{c1}, q_{c2}	
q_{allow}	Allowable bearing pressure.
$Q_{(\infty)}$	Volumetric flow rate at infinite time.
R_f	Friction ratio for Dutch probe.
s	Slope of a curve.
S	Degree of saturation.
SL	Shrinkage limit.
t_{90}	Time required for 90 per cent consolidation.
T	Torque.
T_v	Time factor.
u	Pore pressure.
u_a	Pore air pressure.
u_w	Pore water pressure.
v	Specific volume.
V	Total volume of soil.
V_s	Volume of solid particles.
w	Moisture content.
w_l	The liquid limit.
w_p	The plastic limit.
w_s	The linear shrinkage.
W	Weight of a mass of soil.
x	A reduction factor.
z	A depth.

γ	Bulk density.
γ_d	Dry density.
γ_{sat}	Saturated density.
γ_w	Density of water.
γ_{zx}	Engineer's shear strain.
γ'	Submerged density.
δ_x, δ_z	Changes in dimension.
Δ	A movement.
Δe	Change in void ratio.
Δp	Change in total pressure.
$\Delta p'$	Change in effective pressure.
Δu	Change in pore pressure.
$\Delta X, \Delta Z$	Small dimensions.
$\Delta \varepsilon$	Change in strain.
$\Delta \psi$	Change in applied pressure.
ρ	Settlement.
ρ_c	Consolidation settlement.
σ_z	Normal stress in z-direction.
σ	Normal stress.
σ_h	Horizontal stress.
σ_h'	Horizontal effective stress.
$\sigma_1, \sigma_2, \sigma_3$	Principal stresses.
$\sigma_1', \sigma_2', \sigma_3'$	Principal effective stresses.
τ	Shearing resistance.
τ_{zx}	Shear stress on the z-plane in the x-direction.
ϕ	Angle of shearing resistance.
ϕ'	Angle of shearing resistance in terms of effective stress.
ϕ_r'	Residual angle of shearing resistance.
$\varepsilon_x, \varepsilon_z$	Linear strain.
η	Angle of dilatation.
v	Poisson's ratio.
χ	Pore pressure factor.
ψ	Applied pressure.

Index

372   Index

Transverse (s) waves 337; *see also* Shear
    waves
Trees, effect on buildings 109
Trial holes 121, 141, 144, 159, 161, 175
    profiling procedure in 142
    refusal in 153
Triaxial test 165, 268, 269
Tube-a-manchette grouting 170

Unconfined compressive strength, *see*
    Strength
Unconsolidated sediments 47; *see also*
    Transported soil
Underground services 85, 86
Undermining 159, 244
Undersaturated, *see* Rocks, igneous
Undisturbed sample, *see* Soil samples
Unified Soil Classification 32–41, 83,
    160
Unstable slopes 77
Utilidors 139

Vane shear test, *see* Swedish vane
Vermiculite 57
Vibroflotation 121, 124
Void ratio 9, 36, 112, 116, 118, 265,
    268, 271
    effect of organisms on 65

Voids
    in sedimentary rock 50
    in soil 6
    in soluble rocks 126

Wad 125
Wash-boring 166, 169, 293
Water surplus, annual 118
Water table 77, 85, 89, 97, 105, 108,
    113, 125, 130, 153, 165, 167, 169,
    179, 288, 299, 314, 321
    in boreholes 350
    effect of seasonal fluctuation of 109
    in karst areas 125, 178, 345
    in trial holes 141, 159; *see also* Trial
    holes
Wearing course for roads 206
Weathering 53–9
    classification of, in rock mass 351
Well-points 99, 102
Well-pumping 99, 299
Wenner array 346
Woolwich and Reading beds 161

X-ray diffraction 263

Young's modulus, *see* Modulus, elastic